T0135683

Digital Audio Watermarking

Vom Fachbereich Informatik
der Technischen Universität Darmstadt
genehmigte

Dissertation

zur Erlangung des akademischen Grades
Doktor-Ingenieurs (Dr.-Ing.)

von
Michael Arnold
aus Aschaffenburg

Geburtstag:	31.10.1964
Geburtsort:	Aschaffenburg
Referent:	Prof. Dr.-Ing. José Luis Encarnação
Korreferent:	Prof. Dr. Paolo Nesi
Tag der Einreichung:	26. Februar 2004
Tag der mündlichen Prüfung:	01. April 2004

Darmstädter Dissertationen 2004

Bibliografische Information Der Deutschen Bibliothek

Die Deutsche Bibliothek verzeichnet diese Publikation in der Deutschen Nationalbibliografie; detaillierte bibliografische Daten sind im Internet über http://dnb.ddb.de abrufbar.

ISBN 3-8325-0582-2

Logos Verlag Berlin
Comeniushof, Gubener Str. 47,
10243 Berlin
Tel.: +49 030 42 85 10 90
Fax: +49 030 42 85 10 92
INTERNET: http://www.logos-verlag.de

CONTENTS

LIST OF FIGURES

LIST OF TABLES

Acknowledgments

I would like to thank my advisor, Prof. Dr.-Ing. Dr. h.c. Dr. E.h. José Luis Encarnação for providing the research environment. Furthermore, I would like to thank Prof. P. Paolo Nesi for the valuable hints and detailed proofreading of this dissertation. The development of the audio watermarking algorithm was supported by the WEDELMUSIC project (http://www.wedelmusic.org/), a Research and Development project in the IST programme (IST-1999-10165) of the European Commission which was managed by Prof. Nesi.

Special thanks to the staff at the Security Technology department supporting me in designing, implementing and evaluating the algorithmic concepts. These were — in alphabetical order — Marcus Dickerhof, Dr. Frank Graf, Zongwei Huang, Sebastian Kanka, Stefanie Krusche, Simon Lang, Oliver Lobisch, Jan Mussel, Suresh Peddireddy, Kai Schilz, Dr. Volker Roth, Dr. Stephen Wolthusen, and Armin Zeidler.

Very special thanks to the researchers in the watermarking group consisting of Wolfgang Funk, Martin Schmucker, Michael Voigt, and Dr. Christoph Busch the head of department as well as his predecessor in the department, Dr. Eckhard Koch. The fruitfull discussions about common problems in the watermarking research area have been an important help and guidance during the development and improvement of the algorithm. Moreover I would like to thank the members of the group and especially Dr. Stephen Wolthusen for the valuable aid in proofreading this dissertation.

Abstract

This dissertation describes the design, implementation, and evaluation of an audio watermarking algorithm on the basis of a statistical mathematical model. It shows that even based on contradicting requirements a highly configurable method can be developed suitable for a variety of applications.
The design of the basic audio watermarking method, relying on hypothesis testing, is motivated by a detailed discussion of possible application scenarios and corresponding boundary conditions determining algorithmic parameters. Several extensions of the basic algorithm to enable embedding of n bits of information and multiple non-interfering watermarks are presented as necessary for the applicability of the method in certain applications.
The performance of the watermarking algorithm judged in terms of quality of the watermarked tracks and the robustness of the embedded watermarks has been improved in several ways: To ensure the quality of watermarked audio signals, the integration of a psychoacoustic model is presented, which can easily be adapted to new developments of more sophisticated models. Improvements on the robustness compared to the basic method are demonstrated both on the watermark encoder and detector.
To permit an exhaustive evaluation of the method in terms of quality of the watermarked audio tracks, methods used in the evaluation of high quality lossy compression audio coders have been investigated

and adapted to the watermarking problem. Both subjective listening tests and objective measurement standards have been applied to the watermarked tracks in order to verify the obtained quality, which was validated by a comparison to audio tracks compressed via the MPEG 1 Layer III standard.
A general approach for a risk analysis of watermarking applications is presented, analyzing possible attacks during the deployment of watermarking methods. This is supported by a categorization of attacks on watermarking systems based on their effect during watermark detection and the assumptions made about attackers. According to this classification, the robustness and the security of the watermarks has been demonstrated by a detailed evaluation of the method with regard to removal, de-synchronization, embedding and false alarms attacks.

CHAPTER
ONE

Introduction

While there has been copying of music recordings since technical means became first available and most nations have adopted the concept of "fair use" into their copyright laws, digital means for lossless copying and redistribution pose a serious threat to commercial interests in such recordings of both musicians and the international music industry. This change in available technology has had a noticeable effect on copying for personal use; while the barrier to redistribution previously was the need to copy tracks to analog tape, possibly from multiple sources, resulting in a certain amount of effort required on the part of the copier and – at least to some extent – a loss in quality and convenience compared to the original, this barrier no longer exists since entire CDs can be copied unattended in less than five minutes. Another venue which eliminates even this modest burden on the copier is the use of online digital redistribution using a variety of technologies. Here the copier is burdened only by the bandwidth requirements for distribution; since each copy is identical to the source, the bandwidth for distribution grows exponentially. Given that most of the recipients of such copies appear to accept the degradation in quality inherent in lossy compression mechanisms, the requirements for bandwidth are fairly modest; even residential broadband access permits the copying of a full CD in approximately ten minutes, which can be further ameliorated by scheduling transfers to off hours.

A greater threat, however, emanates from organized piracy. Here, perpetrators will copy not only the digital content but also the physical representation (e.g. CD covers) resulting in – at least at first glance – a faithful copy of the original material. Since CD replication can be done at a massive scale given readily available commercial equipment, this allows the pirates to supply unsuspecting consumers with illegal copies for which the consumer may pay full retail prices. As a result of these developments, content owners are seeking reliable solutions for the problem of protecting their intellectual property rights (IPR) to the recorded material.

1.1 Motivation

The first technologies used for protecting content typically rely on the use of cryptographic techniques. In this scenario the content is encrypted before delivery by means of a secret key which is provided to the customer having purchased the music legally. By the use of encryption techniques the content is transformed in a representation which is useless to the customer unless it can be transformed back requiring the knowledge of the secret key. In turn the fact that these methods rely on the special representation results in a loss of control over the protected data and the unability to further monitor the handling of the content after decryption by a pirate acting as a legitimate client. As a consequence additional security technologies are required not relying on a special representation of the content but complementing traditional protection of the content during transmission already provided by cryptographic mechanisms.
The problems alluded to the unauthorized and uncontrolled redistribution of digital content, justify the development and deployment of mechanisms that permit ex post facto identification of unauthorized copies or, depending on the application scenario, even specific traitor tracing mechanisms.
Digital watermarking of audio data can provide a solution to the identification problem and fill the gap by extending the protection of the content even after decryption by embedding information into

the carrier signal. The embedding of the watermark respectively coupling of the watermark signal and the carrier signal instead of using representations like special formats prevents the removal of the watermarks during normal usage. This dissertation shows that digital audio watermarks can be designed to survive operations including lossy compression, conversions between digital and analog representations, re-encryption, manipulations and attacks.
The copy prevention and copyright protection have been the main motivations of this relatively young research field of watermarking due to the piracy especially in the music sector as stressed by the music labels and presented in the next section. Nevertheless digital watermarking of audio data can be used in different solutions for protecting digital audio data through technical means presented in section 1.3 on page 6 further detailed in chapter 3. Moreover, there is a range of application scenarios beyond that of content protection for which digital watermarks are also very much suitable, particularly for situations where there is no attacker trying to destroy the watermark (see section 3.3 on page 44).

1.2 Piracy In The Music Sector

Piracy in the music sector can be roughly grouped in two sectors: the traditional copying of music in form of tapes, manufactured CDs, CD-R, and audio files distributed and exchanged over the Internet. According to a report of the International Federation of Phonographic Industry (IFPI) [44] the worldwide pirate music market totaled approximately to 1.8 billion units in the year 2000. This is equivalent to a third of all records produced worldwide. The increase of this form of music piracy is driven by

1. the increasing non-volatile storage capacity available to consumers,
2. the growth of the CD-Recordable disc market,
3. and the increases in performance of CD-R burning equipment.

Digital audio signals in CD format typically consist of 16 bit samples recorded at a sampling rate more than twice the actual audio bandwidth (e.g. 44.1 kHz for Compact Discs). Therefore one second of stereo music in CD quality is represented using more than 1.4 Mbit. By using lossy compression methods like MPEG audio coding, the amount of data can be reduced from the CD data rate by a factor of 12 without a significant loss of sound quality for most listeners [73]. Factors of 24 and even more are possible but they reduce the sound quality of the audio track to an extent noticeable to average listeners. The high compression rates are realized by perceptual coding techniques addressing the perception of sound waves by the human ear. The MPEG coding system consists of different layers I-III [1] with increasing encoding complexity and performance[2].Possible data reduction rates using the various layers of MPEG audio are shown in table 1.1. For this purpose a number of freely available tools are

1:4	Layer I	(384 kbps for a stereo signal)
1:6 ... 1:8	Layer II	(256... 192 kbps for a stereo signal)
1:10 ... 1:12	Layer III	(128... 112 kbps for a stereo signal)

Table 1.1: MPEG audio compression ratios

available which permit even individuals with modest skills to readily copy digital audio data from CDs and to make these available to others. Given the performance of available computers, this permits the copying of the digital audio track from CD and simultaneous encoding using MPEG Layer III at the maximum speed permitted by the CD drive without significantly affecting the performance for other tasks apparent to the user.
Due to these effective compression algorithms audio tracks are no longer protected by their once considerable data size. As a result, illegal copying and distribution of music via the Internet must be

[1]The different levels in the coding hierarchy of the audio system defined in [49].

[2]The Audio Layer N decoder is able to decode bit stream data which has been encoded in Layer N and all layers below N.

considered a new form of piracy that was made possible primarily due to the reduced size of the digital audio data and the increased bandwidth available to both consumers and organized crime. This massive distribution potential for pirated copies already appears to severely curtail the market volume of physical CDs, particularly that of CD singles.
One can distinguish between three broad categories of music piracy on the Internet:

WWW and FTP The WWW provides a convenient mechanism for cataloging pirated audio data and can be searched using powerful general purpose search engines as well as using dedicated databases attached to large repositories of pirated music. FTP, which played a similar role for software piracy including the use of search-able distributed indices prior to the WWW is still a widely used for both instant and scheduled bulk data transfers from sites offering pirated audio.

Newsgroups, IRC, and IM NNTP Newsgroups and similar more ethereal services are used to exchange information about new WWW sites with audio tracks to download. Such sites may be run by individuals either willingly or unknowingly violating copyright law or they may have been set up surreptitiously by pirates after compromising an otherwise unsuspecting host. In addition, contacts among different newsgroup users are established to enable a private exchange of such data.

Peer-to-peer networks Peer-to-peer (P2P) services are based on the exchange of data stored locally on the systems of participating users of such services via specialized client software permitting both the retrieval and serving of audio data. Users generally need to register with the service so as to permit downloading of the own data. In turn, the user has access to the music data contributed by all the other users participating in such services. According to the piracy report of IFPI [44] some 2.8 billion songs were traded on Napster in February 2001. An interim injunction was issued against Napster pending further proceedings. According to the injunction Napster had to block its service for files with a possible infringement of copyright. However, given that P2P networking is technologically well understood and easy to implement, the appearance of a number of alternative services such as Gnutella and Freenet have substituted Napster as P2P networks for distributing possibly pirated audio files while Napster went bankrupt.

Different from Napster, some of these services do not rely on centralized server nodes for maintaining user and data registration information. Rather, that information is distributed across all nodes participating in a given network which may be established dynamically and ad hoc. This difference makes it extremely difficult to combat piracy since there is no individual or organization that can be forced by court order to cease operation.

1.3 The Protection of Audio Data through Technical Means

The proposed solutions for eliminating or reducing the losses stemming from piracy can be sorted into several categories:

Copy protection This is the most direct form of control exertion. An entity is sold or licensed for a fixed number of copies; the mechanism must ensure that no additional replication takes place. Several subtaxa can be identified:

Digital physical media Copying the medium must require additional operations in excess of those required for simple reproduction of the work while requiring the presence of the additional information or features for reproduction.

Digital ephemeral data The data required for reproducing the work must be processed by a device which does not permit replication of protected works.

Usage monitoring Each individual instance of an operation or set of operations defined as an usage must be recorded or communicated in such a way that the information can subsequently be used by the rights owner of a work or their agent.

Distribution tracing The creation of a copy and subsequent transmission thereof to another device or individual or the forwarding of the original instance of the work must result in the creation of information recording a feature identifying the source, and may also result in the creation of information recording a feature identifying the destination of the transmission.

Usage control Each individual instance of an operation or set of operations defined as an usage must be subject to the approval of the rights owner of a work or an agent thereof.

The mechanisms that can be employed to reach these objectives range from physical features of media or devices to pose difficulties in copying to elaborate digital rights management schemes tracing the distribution of individual pirated digital copies of works using fingerprinting and digital watermarking techniques.

1.3.1 Copy Protection of Audio CDs

The usage of copy protection mechanisms of Audio CDs as a preventive protection of intellectual property predates by far post facto evidence of violated copyright for rights owners. Nevertheless these mechanisms present a significant danger since a single instance of the protection mechanism failing alone permits, in principle, an indefinite number of identical copies [15].
One example of such a protection mechanism developed by Sony DADC is the Key2Audio family of schemes. Key2Audio realizes its protection by creating a multisession CD with the audio tracks as specified in the so-called "Blue Book" (a format defined by Phillips, Sony, Microsoft, and Apple) in the first session, and a second session containing data. However, the data track is in violation of the Blue

Book standard since the session is not finalized and is inconsistent with the directory structure. The desired result is that the audio session will be played back by audio-only devices while CD-ROM drives attached to computers which can potentially be used for duplication will not be able to recover from errors encountered due to the inconsistent data found in the data session and will not be able to read the audio tracks. The initial Key2Audio schemes were easily defeated by obscuring the (visible) second session, e.g. using a felt pen.

While customers wishing to play back protected audio CDs on computers equipped with CD-ROM devices may employ such circumvention techniques, it appears more likely that the audio data will be distributed in the form of pirated copies once it has been extracted. Digital watermarks can be used to establish copy protection or copy control of audio data in conjunction with special devices able to detect the watermark in order to provide the desired behavior of the recording device or audio player (see section 3.2 on page 42). Even in the event of an unauthorized copy the watermark can assist in distribution tracing (see below) since unlike traditional textual copyright notices these markings are transferred during the copying operation.

1.3.2 Usage Monitoring and Distribution Tracing

Usage monitoring, mainly of interest in the area of multimedia data where monitoring solutions e.g. for broadcasting are well established for the determination of royalty payments based on playlists and the verification of broadcasting of commercials in accordance with contracts is problematic in that most ways of formatting and attaching metadata on the media themselves require standardization of exchange formats, do not survive encoding transitions (particularly to the analog domain) and are easily removed.

Digital watermarking i.e. the embedding of an additional signal within the carrier signal itself and fingerprinting i.e. the derivation of characteristic patterns for identification of media data from the signal itself represent solutions for this application scenario. Both methods are independent of media formats and encodings and can be implemented unobtrusively (see section 3.5 on page 48). Digital watermarks also can assist in the process of distribution tracing if the markings in the media are embedded dynamically during delivering to identify the path in the distribution graph (see section 3.4 on page 46).

1.3.3 Usage Control

Usage control implies that the control mechanism overrides operations counteracting the interests of the user of the system providing the controlled access and usage facility. This is often established by distributing personalized audio players which are designed to create a secured environment within the computing environment of the end user. It is thus a highly intrusive mechanism as it may counteract other property rights such as preventing an individual from reproducing works which are created or owned by that individual. Moreover in the case of usage monitoring mechanisms it has the potential to violate privacy rights and expectations in an egregious manner. In addition, in the case of personalized audio players, the effectiveness of these schemes rely on the tamper-protection of software or hardware devices [10].

1.4 Structure of this Dissertation

This dissertation is divided into nine chapters. The following description provides a brief overview of the structure of the dissertation and the topics addressed in the individual chapters.

Chapter 1 presents the problems related to the protection of digital audio data. Different forms of piracy are presented together with technical approaches to enable the protection of audio data. The defencies of the various attempts are detailed which motivate the development of digital audio watermarking methods.

Chapter 2 clarifies the relation of digital watermarking to the other content protection mechanisms like cryptography and steganography. It provides background information on the terminology used and the formal description of a watermarking system. Furthermore it presents a watermarking model based on communication which lays the groundwork for the following chapters.

Chapter 3 discusses on the basis of preceding chapter applications for digital audio watermarking. Special emphasis is given to a detailed discussion on the requirements concerning the deployment of the watermarking system to these scenarios in order to be effective.

Chapter 4 briefly presents psychoacoustic facts and models used in audio watermarking approaches in order to ensure the inaudibility of the embedded watermarks as one of the main requirements. This presentation is important for a deeper understanding of related research discussed in chapter 5 and the development of the audio watermarking algorithm in this dissertation.

Chapter 5 discusses research related to the subject of this dissertation, particularly in the embedding of the watermarks into the phase, the echo hiding methods, watermarking of compressed audio data and spread spectrum audio watermarking. Additional references to related work are given throughout the remaining chapters where appropriate.

Chapter 6 presents the audio watermarking method developed. The modeling of the method is motivated by the discussion of the requirements and corresponding design criteria. The description of the basic algorithm is followed by the extensions of the method to enable blind detection and the embedding of multiple non-interfering watermarks, which are necessary preconditions for the applicability of the algorithm in different applications. The integration of the psychoacoustic model in the audio watermarking algorithm is of vital importance for the quality of the watermarked audio tracks and the applicability of the method. The robustness of the watermark as one of the main requirements of the watermarking methods is improved by several approaches presented in detail.

Chapter 7 as the first benchmarking chapter evaluates the quality of the tracks watermarked with the algorithm developed in the preceding chapter. The subjective listening tests on the basis of classical psychophysical experiments are introduced and the results are presented. Objective measurement methods of high quality audio data to judge the transparency of the compressed tracks are briefly presented and applied to the watermarking problem. The results from both tests are compared against each other and to MPEG 1 Layer III compressed audio tracks.

Chapter 8 benchmarks the robustness of the watermarks embedded by the watermarking algorithm. First a risk analysis discusses the relevance of attacks in relation to the application in mind. The classification of attacks developed in this chapter is followed by presentation of the metric used in order to quantify the robustness of the embedded watermarks against the attacks presented. Furthermore countermeasures against attacks not related to the signal processing properties of the underlying watermarking algorithm are presented to ensure the applicability of the watermarking method to security related applications.

Chapter 9 summarizes the results obtained and provides an outlook on additional research subjects as well as a critical view of the security obtainable by applying the developed audio watermarking method.

CHAPTER

TWO

Digital Watermarking

Different types of mechanisms have been developed in order to provide protection of multimedia content for the rights owner. This chapter starts by presenting the techniques ranging from cryptography, steganography and digital watermarking and their relationship in section 2.1. This presentation is followed by a discussion of the basic principles of the relatively young field of digital watermarking in section 2.2. This includes a formal description of watermarking as well as of different types of watermarks driven by the specific applications. The systems for embedding and retrieving watermarks are presented from a general point of view and are classified by the information used in the detection process of the digital watermarks by the watermark detector.

Since watermarking is a form of communication with the cover object as the transmission channel, section 2.3 reviews a common model of communication theory and presents the corresponding watermarking communication model with the basic building blocks. Moreover, this model is detailed by presenting the basic and advanced watermark embedder and detector according to the block architecture of the watermarking communication model.

2.1 Cryptography, Steganography and Digital Watermarking

One of the historic applications for digital watermarking is content protection. The following section describes the relation between digital watermarking and other techniques used for content protection. It should be noted, however, that digital watermarking is not limited to this specific application area. Moreover, as shall be discussed later, this is the least suitable application scenario (see section 3.2 on page 42). Most content protection mechanisms rely on cryptological (cryptographical or steganographical) means for the provision of functionality. Cryptographical methods are the most widely used methods for protecting digital content. They provide *confidentiality* by restricting the access to the content to authorized entities. This is done by encrypting the data prior to delivery. The corresponding decryption key is delivered to the entities who legally purchased the digital product. Cryptography provides *active protection* of the digital data during transmission, because the encrypted data is of no use for a pirate without access to the necessary key.

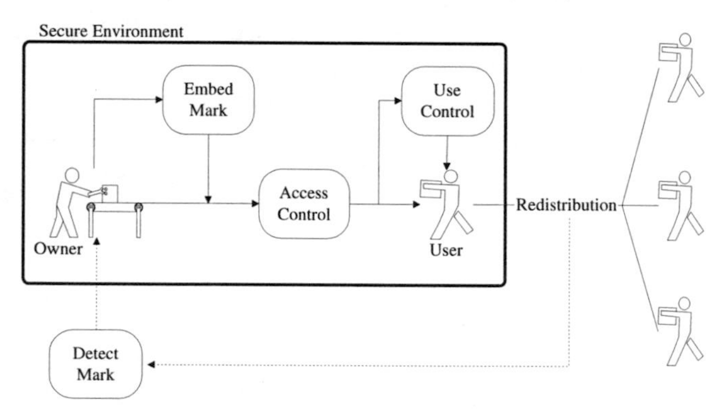

Figure 2.1: Protection scheme including watermarking

The usage of personalized audio players containing the secret key for decrypting the content is one example of securing the data during transmission and the attempt to establish usage control (see section 1.3.3 on page 9). In this case, the audio track is encrypted with a personalized secret key integrated in the audio player, which permits only the playback of the track. However, the control over the audio data is lost if it emerges out of this secured environment. At this stage cryptography does not provide technical means for controlling or monitoring how the content is handled. Therefore, at this point complement protection mechanisms like watermarking techniques can fill the gap by embedding a robust link between owner and content into the content which can be used for a later verification (see figure 2.1 on page 14). Watermarking mechanisms are strongly related to steganographic techniques which are presented in the next section.

2.1.1 Steganography

The distinction between cryptography and steganography was not made ab initio; the term "steganographia" first appears in a manuscript by Johannes Trithemius that was started in 1499 but never completed and did not yet make the distinction between the two terms [102]; this was still the case in a book by Caspar Schott published in 1665 largely containing cipher systems [88]. The narrower definition of cryptography is due to the founder of the Royal Society, John Wilkins who defined the term as *secrecy in writing* [13, 52].
Steganography is the study of techniques for hiding the existence of a secondary message in the presence of a primary message. The primary message is referred to as the *carrier signal* or *carrier message*; the secondary message is referred to as the *payload signal* or *payload message*.
Classical steganography, i.e. steganographic techniques invented prior to the use of digital media for communication can be divided into two areas, *technical steganography* and *linguistic steganography*. The classification of the various steganographic techniques is shown in figure 2.2 on page 16 and described briefly in the following section.

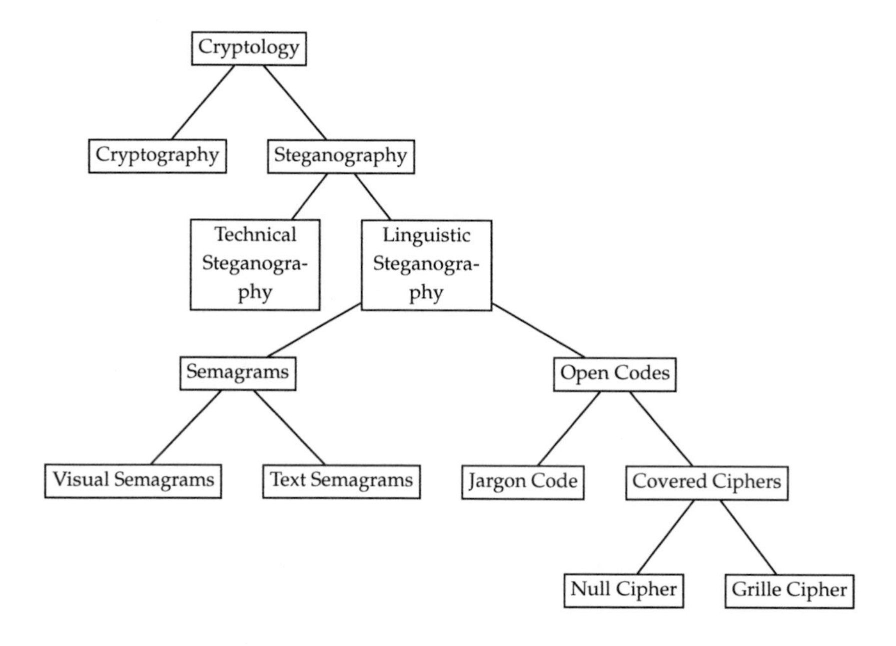

Figure 2.2: Classification of steganographic techniques (adapted from [13])

Steganography itself offers mechanisms for providing confidentiality and deniability; it should be noted that both requirements can also be satisfied solely through cryptographic means.

Technical Steganography

Technical steganography involves the use of technical means to conceal the existence of a message using physical or chemical means to accomplish this goal. Examples of this type of steganography include invisible inks that have been known since antiquity [52, 74] or, more recently, photomechanical reduction resulting in so-called microdots that permit the reduction of a letter-sized page onto an area of photographic film no larger than the dot at the end of this sentence.

Although credit is commonly assigned to German intelligence (who made extensive use of microdots prior to and during World War II), the first documented microdots, albeit some 5000 mm^2 in size, were used by the French during the Franco-Prussian war of 1870-71 and primarily intended for transportation by pigeons [41, 82]. While a fascinating subject in itself, the discussion here is not concerned with it.

Linguistic Steganography

Linguistic steganography itself can be grouped into two categories, *open codes* and *semagrams*. The latter category also encompasses *visual semagrams*. These are physical objects, depictions of objects or other diagrams with an ostensibly innocuous purpose in which a message is encoded. Examples of semagrams include the positioning of figures on a chessboard or the drawing of dancing men in Doyle's "The Adventure of the Dancing Men" [33] shown in figure 2.3.

Figure 2.3: Sherlock Holmes' dancing men semagram

Text semagrams are created by modifying the appearance of a text in such a way that the payload message is encoded. A number of techniques have been devised for this purpose for both manuscripts and printed books. Examples include the interruption of connecting lines in longhand manuscripts, the use of slightly different fonts from the main body of the text to encode the payload message [31, 52], or the punctuation of characters either representing directly or by virtue of another encoding such as the distance between characters using e.g. pinpricks or markings in (invisible) ink over the selected characters [82].
The category of open codes is characterized by embedding the payload signal in such a way in the carrier signal that the carrier signal itself can be seen as a legitimate message from which an observer

may not immediately deduce the existence of the payload signal. The most obvious use of open codes occurs in the use of codewords such as the *Ave Maria* code by Trithemius [103] where individual characters, words, or phrases are mapped onto entities of the carrier signal. Occasionally the use of such codes is unintentional as is made evident by the term commonly used for such open codes, *jargon code*. Conversely, *cue codes* use a possibly elaborate carrier message to signal the occurrence of an event whose semantics have been prearranged. One of the most frequently cited examples is the cue code with which Japanese diplomats were to be notified of impending conflict. In this code, *"HIGASHI NO KAZE AME"* ("east wind, rain") signified pending conflict with the United States while *"KITANO KAZE JUMORI"* ("north wind, cloudy") indicated no conflict with the U.S.S.R. and *"NISHI NO KAZE HARE"* ("west wind, clear") with the British Empire[1] [87]. Unlike jargon codes which lead to atypical language that can be detected by an observer, cue codes are harder to detect provided that their establishment has not been compromised. Another mechanism, commonly referred to as *grille ciphers* is based on the imposition of a grid known only to the communicating parties onto a message consisting of characters or words commonly attributed to Girolamo Cardano and reading the elements left uncovered by the grille in a predefined order [13, 51]. It was still in active use by the German army in 1914 [36].

A variation on the theme of the grille is the use of *null ciphers*. The payload message is embedded as plaintext (hence the null cipher) within the carrier message. The communicating parties must prearrange a set of rules which specify the extraction of the payload message (occasionally also found in literature, an acrostic construct arranges verses in such a way that initial or final letters spell out another word; more elaborate versions were used for steganographic purposes). The payload message may also be subject to encoding prior to embedding in the carrier message; this technique was used

[1]The information containing this code was dispatched on November 26th from the Japanese Foreign ministry to Diplomatic and Consular Officials in enciphered form; while the message could be deciphered by the Navy Department on December 5, 1941, it did not result in adequate countermeasures.

by Johann Sebastian Bach in a number of works; the canonical example here is *"Vor deinem Thron"* (BWV 541) which contains a sequence where g occurs twice, a once, h three times, and c eight times; while this and other isopsephic encodings have been found [94], this has also been the subject of debate [98].

2.1.2 Digital Watermarking

The original purpose of steganographic mechanisms has been information hiding. The techniques and extensions to them based on the possibilities provided by the digital representation of media, however, suggest another application area, namely the protection of a marking against removal. In analogy to a mechanism for the analog, paper domain [54, 27], this was termed *digital watermarking*.
Steganography and watermarking describe methods to embed information transparently into a carrier signal. Steganography is a method that establishes a covered information channel in point-to--point connections, whereas watermarking does not necessarily hide the fact of secret transmission of information from third persons. This implies additional properties not present for steganography since in digital watermarking one must assume that an adversary has knowledge of the fact that communication is taking place. These requirements are that the embedded signal must be redundant so as to provide robustness against selective degradation and removal and it must be embedded in such a way that it cannot be replaced by a fraudulent message or removed entirely; the latter goal is typically achieved by requiring knowledge of a secret for embedding and removal.

2.2 General Principles and Terminology

During the early to mid-1990s digital watermarking attracted the attention of a significant number of researchers after several early works that may also be classified as such [42]. Since then the number of publications has increased exponentially to several hundred per year [27]. It started from simple approaches presenting the basic principles to sophisticated algorithms using results from communication theory and applying them to the watermarking problem.

2.2.1 Basic Principles of Watermarking

Since this research field is still relatively young and has contributors from several disciplines with varying traditions, the terminology used is still quite diverse. This section provides a formal introduction to watermarking systems and the terms used in this context for their presentation.

Formal description of watermarking The basic principle of current watermarking systems are comparable to symmetric encryption as to the use of the same key for encoding and decoding of the watermark. Each watermarking system consists of two subsystems: a watermarking encoder and a decoder. Formally a watermarking system can be described by a tuple $\langle \mathcal{O}, \mathcal{W}, \mathcal{K}, E_K, D_K, C_\tau \rangle$, $\mathcal{O}$ is the set of all original data, $\mathcal{W}$ the set of all watermarks, $\mathcal{K}$ the set of all keys. The functions

$$E_K : \mathcal{O} \times \mathcal{W} \times \mathcal{K} \longrightarrow \mathcal{O} \tag{2.1}$$

$$D_K : \mathcal{O} \times \mathcal{K} \longrightarrow \mathcal{W} \tag{2.2}$$

describe the embedding and detection process, respectively. According to the notation used througout this dissertation the original or carrier object is denoted by $\mathbf{c_o} \in \mathcal{O}$. The watermark is specified by $\mathbf{w} \in \mathcal{W}$ with the extracted and possibly manipulated version $\hat{\mathbf{w}} \in \mathcal{W}$. Correspondingly the watermarked object is $\mathbf{c_w} \in \mathcal{O}$, and the manipulated version $\hat{\mathbf{c}}_\mathbf{w} \in \mathcal{O}$. The comparator function

$$C_\tau(\hat{\mathbf{w}}, \mathbf{w}) : \mathcal{W}^2 \longrightarrow \{0, 1\} \tag{2.3}$$

compares the extracted with the acutally embedded watermark using the threshold τ for comparison. The input parameters of the embedding process are the carrier object (or original $\mathbf{c_o}$), the watermark $\mathbf{w}$ to be embedded, as well as a secret or public key $K \in \mathcal{K}$:

$$E_K(\mathbf{c_o}, \mathbf{w}) = \mathbf{c_w} \tag{2.4}$$

The output of the encoder forms the marked data set (see figure 2.4). In the detection process, the marked and possibly manipu-

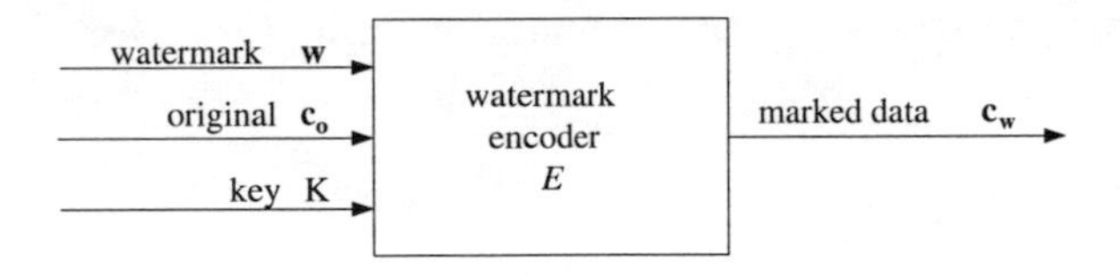

Figure 2.4: Generic watermark encoder

lated data set $\hat{\mathbf{c}}_\mathbf{w}$, the original $\mathbf{c_o}$, the watermark $\mathbf{w}$, and the key K used during the embedding process are the input parameters (see figure 2.5). The various types of watermarking systems differ in

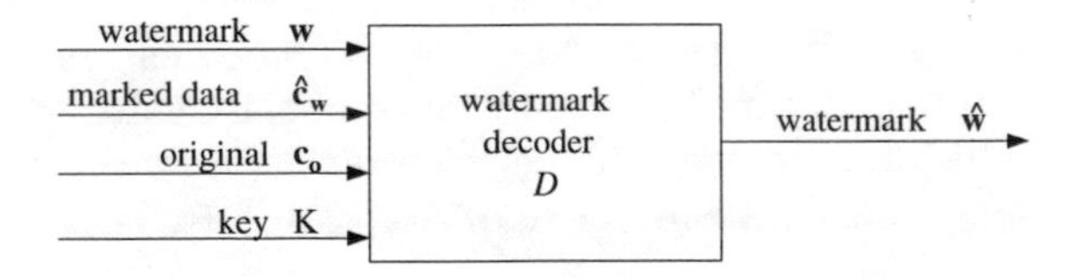

Figure 2.5: Generic watermark decoder

the number of input parameters in the reading process (see section 2.2.2 below). The extracted watermark $\hat{\mathbf{w}}$ differs in general from the embedded watermark $\mathbf{w}$ due to possible manipulations. In order to judge the correspondence of both watermarks the comparator function C_τ compares the suspected watermark with the retrieved one against a threshold τ:

$$C_{\tau}(\hat{\mathbf{w}}, \mathbf{w}) = \begin{cases} 1, & c_{\tau} \geq \tau \\ 0, & c_{\tau} < \tau \end{cases} \tag{2.5}$$

The threshold τ depends on the chosen algorithm and should in a perfect system be able to clearly identify the watermarks. This formal analysis of the watermarking systems can also be used to develop a geometric interpretation of the watermarking algorithms as shown in [27].

2.2.2 Terminology

Types of watermarks

- *Robust watermarks* are designed to resist against heterogeneous manipulations; all applications presupposing security of the watermarking system require this type of watermark. In turn the security of the watermarking system is defined by the possible attacks in the specific application scenarios and corresponding robustness requirements (see section 8.1 on page 148).

- *Fragile watermarks* are embedded with very low robustness. Accordingly, this type of watermarks can be destroyed even by the slightest manipulations. In this sense, they are comparable to the hidden messages in steganographic methods. They can be used to check the integrity of cover objects.

- *Public* and *private watermarks* are differentiated in accordance with the secrecy requirements of the key used to embed and retrieve markings. According to the basic principle of watermarking the same key is used in the symmetric encoding and decoding process. If the key is known this type of watermark is referred to as public, if the key is hidden as private watermarks. Public watermarks can be used in applications which do not have security-relevant requirements, e.g. for the embedding of meta-information.

- *Visible* or *localized watermarks* can be logos or overlay images in the field of image or video watermarking. Due to the implicit localization of the information these watermarks are not robust.

Besides the various types of watermarks, four different watermarking systems are classified according to the input and output during the detection process. Using more information at the detector site, on one hand, increases the reliability of the whole watermarking system but limits the practicability of the watermarking approach, on the other hand.
The side information in the detection process can be the original $\mathbf{c_o}$ and the watermark $\mathbf{w}$ itself (see figure 2.5 on page 21). Accordingly four permutations of side information requirements are possible.

Watermarking systems

- *Non-blind watermarking* [2] systems require at least the original data in the reading process. Furthermore one can further subdivide this type of system depending on whether or not the watermark is needed within the decoding process.

  Type I systems detect the watermark of the potentially manipulated data set by means of the original:

$$D_K(\hat{\mathbf{c}}_{\mathbf{w}}, \mathbf{c_o}) = \hat{\mathbf{w}} \qquad (2.6)$$

  Type II systems additionally use the watermark and therefore represent the most general case:

$$D_K(\hat{\mathbf{c}}_{\mathbf{w}}, \mathbf{c_o}, \mathbf{w}) = \hat{\mathbf{w}} \quad \text{and} \quad C_\tau(\hat{\mathbf{w}}, \mathbf{w}) = \begin{cases} 1, & c_\tau \geq \tau \\ 0, & c_\tau < \tau \end{cases} \qquad (2.7)$$

  These systems answer the question: Is the watermark $\mathbf{w}$ embedded in the dataset $\hat{\mathbf{c}}_{\mathbf{w}}$? In this way the information content of the watermark is one bit. By using this information the robustness of these watermarking methods is in general increased.

[2]The term *private watermarking* is also used, which can lead to confusion due to the previous term of public and private watermarks.

- Different from the above method, *semi-blind watermarking* does not use the original for detection:

$$D_K(\hat{\mathbf{c}}_\mathbf{w}, \mathbf{w}) = \hat{\mathbf{w}} \quad \text{and} \quad C_\tau(\hat{\mathbf{w}}, \mathbf{w}) = \begin{cases} 1, & c_\tau \geq \tau \\ 0, & c_\tau < \tau \end{cases} \tag{2.8}$$

  This is essential in applications where access to the original is not practical or possible. Semi-blind watermarking methods can be used for copy control and copyright protection.

- *Blind watermarking* [3] is the biggest challenge to the development of a watermarking system. Neither the original nor the watermark are used in the decoding process:

$$D_K(\hat{\mathbf{c}}_\mathbf{w}) = \hat{\mathbf{w}} \tag{2.9}$$

  This is necessary in applications in which n bits of information must be read out of the marked data set $\hat{\mathbf{c}}_\mathbf{w}$ as e.g. during the pursuit of illegally distributed copies.

2.3 Watermarking as a Form of Communication

Watermarking can be considered as communication of the watermark over a channel consisting of the original work to be watermarked. Therefore a natural approach in development of conceptual models for watermarking is to study the similarities between communication models and corresponding watermarking algorithms. Both models transmit data from an information source (the watermark) to a destination (the user or an other system).
The typical model of communication consists of several blocks shown in figure 2.6 on page 25. This model was introduced by Shannon in

[3]Also called *public* or *oblivious* watermarking.

his landmark paper in 1948 [90]. The source message m is transformed via a source encoder into a sequence of binary digits u representing the encoded source as an information sequence. The process is performed in order to minimize the number of bits to represent the source output and to enable the non-ambiguous reconstruction of the source from the information sequence [66].

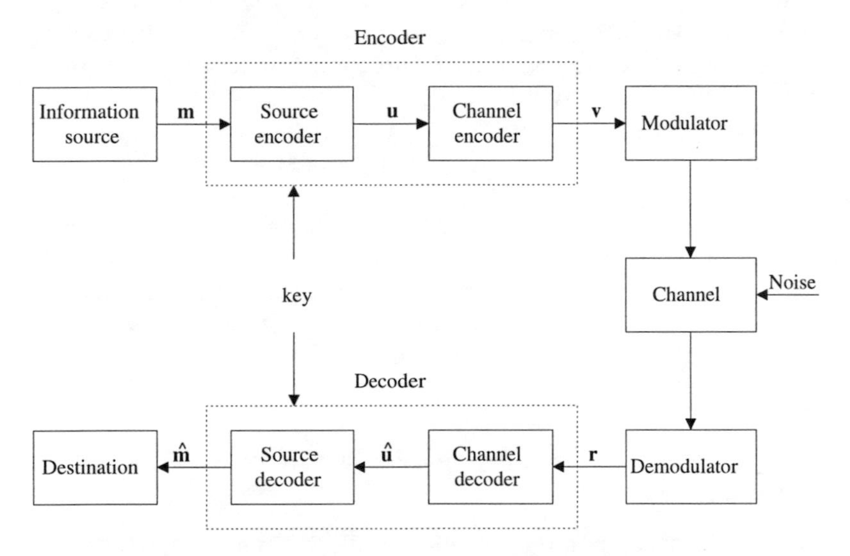

Figure 2.6: Basic communication model for secure transmission

The *channel encoder* transforms the information sequence u into a *encoded sequence v* called a *code word*.
In order to transmit the discrete symbols over a physical channel a *modulator* transforms each symbol of the encoded sequence v into a form suitable for transmission [85].
During transmission over the channel the transformed sequence is distorted by noise. The different forms of noise which can disturb the transmission are driven by the channel characteristics.
On the receiver site the *demodulator* processes the transmitted se-

quence and produces an output r consisting of the counterpart of the encoded sequence. Corresponding to the encoder the *channel decoder* transforms the output of the demodulator into a binary sequence $\hat{u}$ which is an estimation of the true sequence being transmitted. In a perfect channel the estimated sequence $\hat{u}$ would be a copy of the true sequence u. Carefully designed source encoders can reduce coding errors originated from the disturbance by the noise of the channel [66]. The last step is performed by the *source decoder*, which transforms the decoded sequence $\hat{u}$ into an estimate $\hat{m}$ of the source output which is sent to the destination.
The different types of communication channels can be categorized by the type of noise introduced during the transmission and how the noise is applied to the signal [85].
Besides the channel characteristics, the transmission can be further classified according to the security it provides against active attacks trying to disable communication and against passive attacks trying to monitor the communication (read the transferred messages).
The defense against the attacks is based on

- spread spectrum techniques trying to prevent active attacks;
- cryptography encrypting the messages in order to ensure privacy.

Digital watermarking and spread spectrum techniques try to fulfill similar security requirements in preventing active attacks like jamming the communication between different communicating parties. Spread spectrum technologies establish secrecy of communication by performing modulation according to a secret key in the channel encoder and decoder (see figure 2.6 on page 25).
A watermarking model based on communication consists of the same basic blocks like the communication model with different interpretations. There is a direct correspondence between the watermark embedder/detector and the channel encoder/decoder – including the modulation/demodulation blocks – respectively. The message to be transmitted is the watermark itself. The additional requirement of secure transmission of the signal over the channel requires the usage

of a secret key in the encoding and decoding procedure (see section 2.2.1).
The channel characteristics can be modeled by

- the cover object representing the channel carrying the watermark;
- the kind of noise introduced modeled by the different processing that may happen during transmission of the watermarked object. This additional processing may be anticipated manipulations or deliberate attacks.

The encoding block of the watermark embedder encodes the watermark message m into a coded sequence v. During the modulation the sequence v is transformed into a physical signal, the watermark signal w, that can be transmitted over the channel.

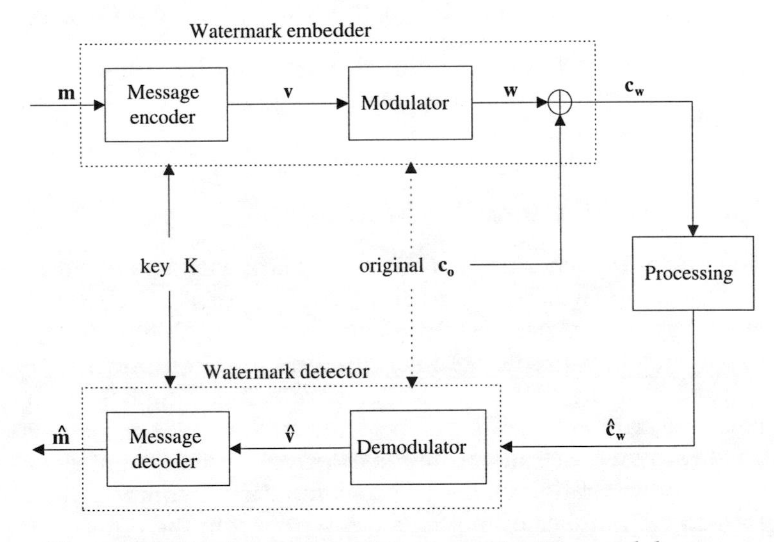

Figure 2.7: Basic watermark communication model

The difference between the marked and original cover object – which actually forms the added watermark – will essentially have the same digital representation as the original data set. For example, in case of an audio file the added watermark will be a signal with the same sample rate and bit resolution as the cover signal. At the watermark detector site the possible, distorted marked object is demodulated into $\hat{\mathbf{v}}$, which is a distorted version of the coded sequence v. The watermark message $\hat{\mathbf{m}}$ is obtained by means of the watermark message decoder (see figure 2.7 on page 27) from $\hat{\mathbf{v}}$.
In analogy to the basic communication system the encoder has to perform three steps

1. Encode the message into a coded sequence using a secret key.
2. Modulation of the coded sequence into a physical representation according to the channel respectively the cover object.
3. Addition of the modulated sequence to the cover object to produce the watermarked object.

In order to classify existing digital watermarking systems, the basic building blocks of watermark embedders/detectors must be examined in closer detail.

2.3.1 First Generation Methods

The first approaches implement watermark encoders by addition of the generated watermark pattern without considering the channel characteristics respectively the cover object c_0. Methods from this first generation predefine a set of significant components in a so-called *embedding domain* for watermark embedding based on some heuristic criteria [25, 14]. As depicted in figure 2.8 on page 29, this usually involves a transformation into another signal representation[4] where the alteration of the pre-selected carrier components is performed. A number of methods are working in the Fourier domain

[4]Typical embedding domains used for modification are the Fourier, wavelet and Cepstrum domains.

using the low to middle frequency range as components to embed the watermark [59].

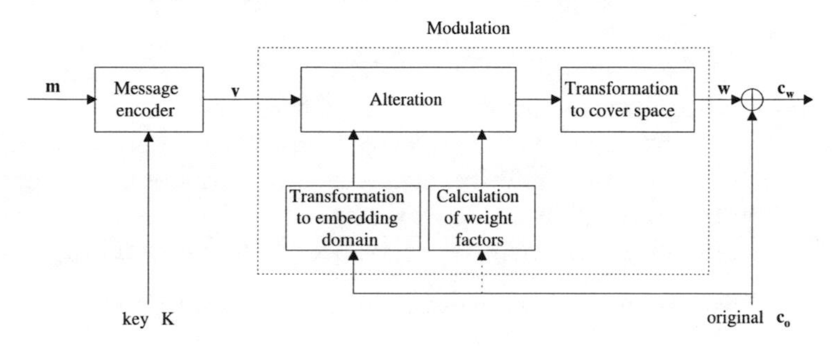

Figure 2.8: Basic watermark embedder

From experiments it is known that these are perceptually significant components [100, 69], resulting in a compromise between quality of the watermarked object as well as robustness of the embedded watermark. To enable the adjustment between the two opposite requirements of perceptual visibility and robustness of the watermark, a vector $\boldsymbol{\alpha}$ of weight factors is calculated. The very first algorithms [25] have used a vector with equal elements

$$\boldsymbol{\alpha} = \{\boldsymbol{\alpha}[i]\}_{i=1}^{N}, \boldsymbol{\alpha}[i] = \alpha_{const}, \forall i \tag{2.10}$$

to control the embedding strength respectively the power of the embedded watermark (see figure 2.8). The obvious disadvantage of such simple methods is the missing relation between the cover object and the embedded watermark resulting in a possible loss of fidelity if an overall weight factor is used not considering local variations of the cover object respectively the channel. Furthermore, optimization of the robustness of the embedded watermark with regard to the actual cover object is not taken into account. From this point of view little or no information about the specific cover object is taken into account to improve both the embedding and detection procedure. More advanced algorithms [97, 72, 11, 81, 8] optimize the quality by

investigating the cover object to calculate perceptual thresholds and the corresponding weight vector $\boldsymbol{\alpha}$.
These so-called perceptual thresholds are derived from perceptual models [105, 92, 76] for different media types. This information is used in the modulation block of the encoder to shape the added pattern to ensure maximum quality. Correspondingly, the added pattern is a function of the cover object. These methods use the perceptual thresholds to optimize for quality usually neglecting the effect on the robustness of the watermark.
The decoders perform two steps to retrieve the watermark:

1. The obtained signal is demodulated to obtain a message pattern.

2. The message pattern is decoded with the decoding key in order to retrieve the embedded watermark.

The demodulation can be performed in different ways. If the original object is available to the decoder, it can be subtracted from received signal which is the watermarked object to obtain the received noisy watermark pattern (see figure 2.9).

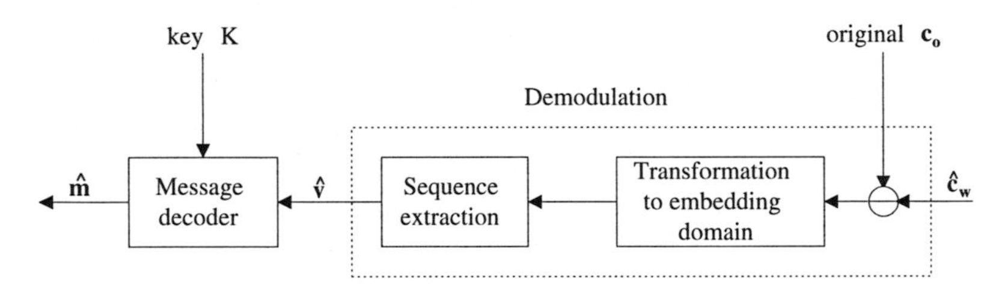

Figure 2.9: Basic watermark detector

Demodulation in this way is done in non-blind watermarking systems (see section 2.2.2 on page 23). Other approaches use data reduction functions to cancel out the effect of the addition of the cover object. This can be done by approximating the original in the detection procedure before subtracting it from the watermarked object.

In the decoding procedure the coded sequence $\hat{\mathbf{v}}$ of the watermark has to be extracted by the sequence extractor. In early approaches the same predefined carrier components in the embedding space are used for sequence retrieval as in the embedding step. The watermark message $\hat{\mathbf{m}}$ is decoded in the watermark message decoder from $\hat{\mathbf{v}}$ by means of the secret key.

2.3.2 Beyond the First Generation

In contrast to the initial approaches, more advanced algorithms take additional information into account in the encoding process. The side information used in the watermark embedder concerns the channel characteristics respectively the cover object $\mathbf{c_0}$. The cover information can be used at several stages of the embedding process. The pre-selection of the carrier elements is not specified in advance according to some heuristic criteria driven by basic perceptual facts using a fixed frequency range but rather by different algorithmic requirements. A feature extraction block is often combined with the transformation in the embedding domain to select carrier components which are more appropriate for watermark embedding than others.
Selection criteria are e.g. reduced correlation between the cover object and the watermark signal to be embedded, or higher possible alterations of components according to the perceptual thresholds to improve the robustness. This information can be used in the alteration block in figure 2.10 on page 32 to optimize between quality on the one hand and robustness on the other hand according to application requirements.
Moreover, the information of the cover object can be used in order to adjust the coded sequence with respect to the message carrier respectively the cover object. A search through a set of code vectors can be performed in order to embed the code vector which is closest to the cover object. The set of all code vectors is tested at the decoder site in order to identify the message corresponding to the code vector found. Research in using side information (the cover object) in the

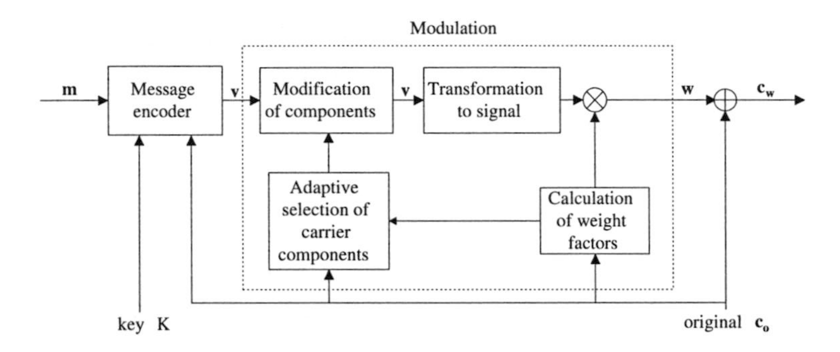

Figure 2.10: Advanced watermark embedder

watermark encoder includes the so-called *dirty paper* channel studied by Costa [24].
By using perceptual models and feature extraction procedures at the watermark detector site, the detection reliability of the watermark can also be improved.
Feature extraction has to be performed in order to select the corresponding regions or components used during watermark embedding. Perceptual models can be applied at the detector site to calculate the perceptual weight factors. Depending on the input of the model which can be the original or the marked data set, the weight factors are exactly the same as in the embedding process or approximations thereof. Since one of the main requirements of watermarking is perceptual transparency of the marked data set one can assume that the calculation of the masking threshold leads to similar results as in the original case. Therefore the approximations of the weight factors derived from the marked cover object should lead to satisfactory results. To improve detection performance, knowledge about the perceptual characteristics can be used by applying inverse weighting to the carrier components in order to remove the perceptual shaping which can be regarded as a noisy process during embedding (see figure 2.11 on page 33).
Besides the basic methods used for modulation of the information

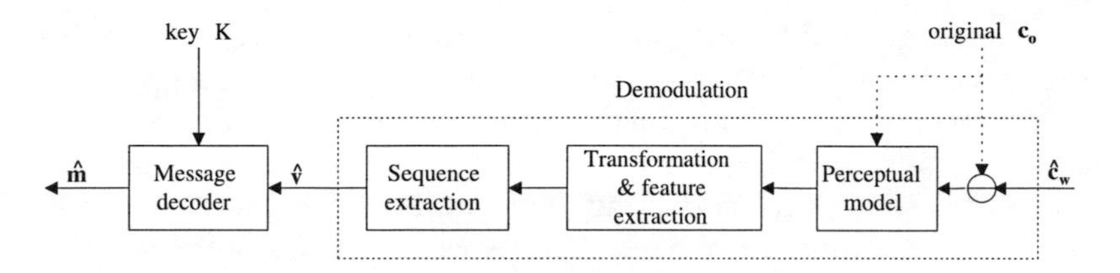

Figure 2.11: Advanced watermark detector

taking the cover object or the transmission channel, respectively, into account to achieve a balance between quality of the watermarked object and robustness of the embedded watermark requires a deep understanding of the characteristics of the carrier data type. Therefore the algorithms for digital watermarking have to be tailored to the specific media type.

2.4 Summary

This chapter has laid the groundwork for the following chapters by starting with the description of digital watermarking from a communication-theortic point of view at the beginning. The classical approaches for digital content protection, namely cryptography and steganography were presented shortly in order to describe their relation to digital watermarking as a special form of steganography. Digital watermarking - in contrast to steganography - has the additional requirement of robustness due to the knowledge of a potential attacker about the communication and the possibility of alterations during standard processing steps.
Having specified the relation to the other content protection mechanisms, the generic building blocks of a watermarking system consisting of a watermark embedder and detector, which are similar to symmetric encryption in using a secret key for embedding and detection, were presented. The terminology concerning the different watermarks differentiates between robust and fragile with respect to

the robustness, public and private according to the secrecy requirements, visible and localized corresponding to their localization. The watermarking systems are further classified with respect to the information available at the watermark detector. They are denoted as non-blind, semi-blind and blind watermarking methods with less information on the one hand and higher complexity with regard to the development on the other hand. Furthermore, watermarking was presented as a form of communication with different building blocks with regard to the basic communication model. This enabled the distinction in first and advanced methods related to the information about the transmission channel respectively the cover object taking into account into the watermark embedder.

CHAPTER

THREE

Applications of Digital Watermarking for Audio Data

From a fundamental point of view the digital watermark establishes a link between the raw audio data and corresponding information in form of a mark which can serve different purposes according to the intended application. The main difference between a marking technique as watermarking in contrast to other labelling mechanisms are the basic features provided by this link: a watermark is not audible whereas a background jingle in the case of audio might be disturbing. Furthermore the watermark is directly embedded in the raw data. This ensures the independence of the watermark from specific data representations, such as different file formats.
Several applications are described in this chapter taking advantage of the features offered by the watermarking technology. The discussion includes advantages by applying the watermarking technology as well as the requirements which have to be fulfilled. Section 3.1 presents the copyright protection application (being the historical motivation for developing watermarking technologies). One of the most important goals of the music industry is to develop a copy protection mechanism to avoid the uncontrolled spreading of their products. The application of digital watermarking to this problem is discussed in section 3.2. Scenarios using the watermark primarily as side channel are presented in section 3.3. These annotations can

be used in a variety of ways resulting in the transaction tracking applications discussed in section 3.4 and the generation of playlists presented in section 3.5.1.

3.1 Copyright Protection

Using the watermarking technology for copyright protection was presumably the primary motivation for the earliest developments. In this case, the watermark is used in order to establish a reliable link between its origin (respectively the copyright owner) and the audio material. On the other hand, copyright notices are used today to provide the evidence of copyright for the rights holder. In the case of sound recordings, they have to take a special form. Furthermore, the copyright notices have to be placed as a label on the physical medium (for example on a CD) or packaging (for example on the cover of the CD). These textual notices come along with two main intrinsic disadvantages: During the copying of the audio material the copyright notice gets lost. Furthermore, audio material cannot be annotated in such a simple way. A perceptible marking can be distracting and would have to be placed in perceptual significant locations [25] since otherwise a simple filtering operation would remove the marking.

3.1.1 Identification of Origin

Digital audio watermarking in turn can provide the requisite embedded information to identify owner by replacing the textual copyright notice. Watermarks are superior to the use of a notice because they are not audible and inseparable from the audio track. The first feature enables the integration of the copyright notices into the audio content itself in contrast to a label on a specific physical representation like a tape, record, CD, SVCD, SACD etc. Furthermore, the inseparableness of the watermark from the audio content ensures the transfer of the copyright notice during the copying. In the most trivial form this amounts to embedding a source-encoded textual notice, but given the limited bandwith afforded particularly by robust digital watermarks, more efficient encoding schemes are called for. Generally, this application scenario requires that information serving as evidence of the origin of a title is provided as the payload signal.

The first requirement of watermarks corresponds to the quality of the signal to which digital watermarks have been applied; presumably the most common quality metric is the absence of perceptually significant artifacts introduced by markings. Besides copying a more realistic application scenario must recognize that the audio track may be subjected to a number of transformations including format conversions. Even for the case of duplication of the entire audio material without alteration, a conversion in the representation requires that the digital watermarks are bound to the actual carrier signal of the creation.

Many transformations to which an audio track may be exposed either deliberately or unwittingly are, however, not lossless. Such transformations may include the use of lossy compression like MPEG 1 Layer III to reduce storage and bandwidth requirements, digital-to-analog conversions, or complex operations such as multiband compression. Nevertheless, the quality degradation due to a variety of transformations inherent in a digital-to-analog conversion may be acceptable as made evident by [44]. It is primarily due to these considerations that the robustness requirement of digital watermarks for copyright protection was introduced. However robustness requirements, particularly with regard to lossy compression, are modest if the end result is to be of value for a potential user.

3.1.2 Proof of Ownership

Besides providing a mechanism to identify the owner the usage of digital watermarking technology to protect the copyright can be extended to the next level of actually proving ownership. This corresponds to the reliability or security of the link between the copyright

owner and the audio material, which cannot be provided by a textual copyright notice. In this scenario, the rights owner will be authenticated by the knowledge of the secret key used for embedding the watermark to extract the owner identification information.
In addition, to be generally applicable for the copyright protection problem the watermarking system must ensure that the false positive rate is below a certain threshold. Depending on the application sub-scenario this can mean the probability that a digital watermark detector responds without a marking being present or that an existing marking is read, but the payload is not retrieved correctly in such a way that another syntactically correct payload is detected (see section 8.3.6 on page 184). The transgression of the threshold would lead to increased expenses due to litigation, warranty claims, or other customer dissatisfaction. In a consumer-oriented application area this imposes a significant burden of proof on the digital watermarking algorithm (e.g. the DVD CCA requires a false positive rate below 10^{-12} [34]).
To protect against unauthorized use or duplication one has to consider the different possibilities of copyright infringement. In general, violations of the intellectual property rights (IPR) can happen at every point of the distribution chain of the audio material. This includes the content provider serving the audio data as well as the end user buying or licensing the material. Protective measures can be grouped into two broad categories. The first category encompasses the protection against misappropriation of creations by other content providers without the permission or compensation of the rights owner, while the second category includes protection mechanisms against illicit use by end users. While both categories are commonly referred to as piracy, the issues involved and requirements for protective measures may make a distinction beneficial, although in a number of application scenarios protection against both categories of misuse are called for.

Misappropriation by Content Providers Misappropriation by other content providers can, in turn, occur in several forms. In the simplest

case, a creation is duplicated, redistributed, or re-sold in its entirety and in its original form. While the audio file may be also subject to quality-lowering transformations the application scenario dictates that the quality cannot degrade to a significant extent. However, a high degree of robustness against desynchronization attacks is desirable as such an attack does not necessarily degrade the perceived quality (e.g., slowly varying warp of the time axes see section 8.3.4 on page 180).
Besides duplicating and redistributing the entire audio track, new compositions using parts (samples) of one or more audio songs can be performed with comparable ease and modest tools. Such compositions can take the form of collections or fragments of existing creations, which may occur over multiple generations. If only one of the intermediate generations omits the required copyright notice, it becomes increasingly difficult to associate the fragment with the proper rights owner. While such composition may be legitimate in the case of very limited fragments under the fair use doctrine, the compensation of the rights owners cannot be avoided in general. In addition, the rights owner may object to having a creation used in a certain context and refuse to grant permission for such uses (e.g. pasting the chorus from one composition into another).
A digital watermark identifying the creator directly or indirectly therefore is beneficial not only to the rights owner but also to legitimate users as it can significantly reduce the effort required to track down all rights owner and may make certain kinds of compositions possible that could otherwise not have been considered.
Such applications add further constraints on the capabilities required for audio watermarking techniques in that the watermark must be retrievable after extensive temporal cropping of the audio material. However, the watermark payload must still be recoverable from the cropped elements, e.g. by repeating the watermark over the audio signal. This implies that the bandwidth requirement for the watermark payload applies not to the entire signal but rather to the minimum fragment size[1] for which recognition must be possible.

[1] Also called minimum watermark segment (MWS).

Illicit use by End Users The main distinction between the unauthorized reproduction and use of creations by content providers discussed in the previous section and unauthorized use by end users lies in the visibility of the perpetrator. Whereas a content provider ultimately must attempt to sell or otherwise profit the creations and therefore must expose himself to the public (albeit possibly in a different jurisdiction), the same is not true for end users.

End users may duplicate creations and distribute such material either within their personal environment or e.g. using file sharing services, some of which can provide a certain amount of anonymization. While copyright markings and digital watermarks such as those discussed in the preceding section can assist in identifying the audio material if and when it is located, identification of the source of the material cannot be accomplished by using these markings.

A deterrence effect may, however, be conjectured if an individual copy of a track is tied to a specific transaction (which may implicitly be extended to an individual based on the type of records maintained for a transaction). The payload for such a digital watermark may be the identity of the end user to which a creation is sold or otherwise licensed or a unique identification of the transaction.

This can lead to the identification of the original purchaser or licensee (or the last authorized link in a distribution chain in case of what has occasionally been called "superdistribution") if a copy or elements thereof are found in the posession of an unauthorized end user.

The payload size required for this application scenario mirrors those discussed in the preceding section; for transaction watermarks the uniqueness constraint must be balanced against the drawbacks of large payloads, at least to an extent that the probability of duplicate transaction identifiers is comparable to other types of false positives in the detection stage since otherwise the evidentiary value could be called into question.

Pragmatic issues also must be taken into consideration in determining the true deterrence effect of marked creations since – unlike in the case of commercial interests – in most situations there will exist a strong legal protection of individuals from searches of an individual's property and invading the privacy of an individual without a viable justification. This is likely to limit the deterrent value in that only copies that are found in the open (e.g. those traded openly by an end user) can be verified for containing watermarks. It is also untested whether such a proof in the form of a digital watermark is sufficient to show that a transgression has taken place – and even so it may be the case that an end user thus identified can plausibly deny the deliberate dissemination.

An argument for the efficacy of digital watermarks as a deterrent in this context is the use of automatic search engines that scan for protected creations. Such searches can occur either by transferring suspicious content to a central location and analyzing the data there, or by using so-called agents to have the analysis process take place in situ [107].

In the case of the application scenario discussed in the preceding section, central analysis is already exceedingly difficult from a logistical standpoint, as the set of potential sources for redistribution is not properly bounded and may grow faster than the product of bandwidth and processing power available to the rights owner. The approach becomes even less attractive when arbitrary end user systems are considered – these will generally not make creations available to external nodes and may also not be available at the time of checking for protected creations [1].

Using agents to detect misuse could, under very benign circumstances, eliminate the logistical problem in the scenario considering misappropriation by other content providers – although making processing power available to rights owners through content providers would almost certainly require commensurate reimbursement, but would leave the problem of the ill-defined set unsolved. It is, however, unlikely that any end user would consent to rights owners executing arbitrary code and granting access to any data located on the end user's system – even though a comparable approach has been proposed in a similar context for a digital rights management system [65]. The robustness requirements for digital audio watermarks protecting against end users are considerably higher, since the quality aspect appears to be of lesser significance if creations are obtained for free or for the cost of transferring the data. This assertion is supported by the observed popularity of highly compressed audio tracks with bitrates up to 64 kbit/s with a quality that is significantly below the level of the original (see table 4.1 on page 55). Given such low quality requirements, robustness must also be maintained against a number of deliberate attacks (as discussed in chapter 8). This is particularly problematic since there exist automatic tools for performing – typically highly successful – attacks against digital watermarking systems (e.g. used in developing benchmarks, or in the course of academic research) that can be used even by individuals with modest skills that introduce quality degradations comparable or in most cases significantly less than those tolerated by end users in the case of compression [30].

3.2 Copy Protection

Digital watermarks can be considered protection techniques only in that they provide a deterrence mechanism or a evidence of breach of copyright or other contractual obligations after the fact.
On the other hand copy protection or control are mechanisms that rather providing an evidence of an copyright infringement are trying to prevent the duplication of illegal copies (see section 1.3.1 on page 7). The protection of audio material can be established during the recording and playback stages. Watermarks can be used for copy protection in conjunction with specially equipped devices for playback and recording. In each case both the presence of a potentially specific marking or the absence thereof can be used to induce a desired behavior of the recording device or audio player controlling the operation. Although this will be hard to establish politically and practically (see section 9.2 on page 190).

Personalized copy A functionality often referred to as *playback control* requires the presence of a watermark to permit playback of audio material. This implies the personalization of the audio tracks for a given individual or set of devices (e.g. personalized audio players) since otherwise a successful duplication of the digital representation would also reproduce the watermark; depending on the robustness requirements for the watermark watermark recognition may even be possible from copies generated from analog sources (e.g. audio signals captured and re-digitized from an analog output of a legitimate playback device). To implement the scenario a watermark detector has to be a integral part of the audio player, which prohibits the playback if the correct watermark is not detected.

Personalization of audio tracks via watermarks is equivalent to the identification of users (or devices) by the transaction watermark. Therefore the requirements for payload capacity match those discussed in section 3.1.2 on page 40. On the other hand watermarks used to identify individual transaction records require a correspondingly higher payload size. As copies must be individually marked in this application scenario, this imposes limitations on the distribution forms that can be justified economically.

Playing only the original As opposed to requiring the presence, the absence of a watermark could be used in an application scenario in which an audio player embeds a watermark. The watermark to be embedded can identify the audio track as it is played back – alternatively this watermark could also be embedded in a recording process – to identify first from subsequent generations of copies. A functionality referred to as *record control*. This is the application scenario most similar to the "copy bit" approach found in the DAT system, along with the familiar threats and vulnerabilities from that approach – albeit with an increase in difficulty if the embedding process is an integral part of decoding an audio song for playback.

Copy Bit implementation Conversely, another application scenario consists in requiring the absence of a digital watermark for the recording stage. Therefore a necessary precondition for this application is that every recording device is equipped with a watermark detector. The benefit compared to a simple "copy bit" mechanism is that removal of the marking once it has occurred requires more effort than would be the case of a marking that is not tied to the content itself. In addition, more elaborate schemes can use the fact that larger payload sizes than a single bit can encode (assuming safeguards against unauthorized manipulation of the payload or the marking itself) arbitrary instructions as to the admissibility of copying or playback operations that can be changed dynamically either in case of duplication or – provided a writable representation – during the use of the representation, e.g. to record the number of remaining playback operations for a given license. Such a mark-on-use scheme was used for the output of the DiVX devices for the playback of digital movies, although the digital watermarks were embedded only in the analog signal and not used for copy protection as such.

3.3 Watermarking as Annotation Mechanism

Although the need for the development of a copyright protection system formed the basic motivation digital watermarks posess a number of desirable properties that make their use in other application scenarios desirable, some of which are analogous to the ones discussed in the preceding sections.
Annotation watermarks [14] provide information in a side channel that is coupled to the carrier signal without degrading the perceived quality of the carrier signal; this distinguishes them from markings that are either perceptible or not tied to the carrier signal (e.g. header fields specific to a certain file format). They are therefore of interest in applications where the format of multimedia data cannot be guaranteed or is likely to change throughout a work flow; similarly, since digital/analog conversions are part of the expected transformations audio data must resist, the properties of digital watermarks are desirable.

Unlike perceptible markings, digital watermarks can also be distributed across an entire audio stream such that the resulting signal can be cropped significantly and the digital watermark can still be recovered either in its entirety or to a significant extent from a fragment (see section 3.1.2 on page 39).

The annotation application scenario can be considered notably distinct from steganographic techniques in that the presence of the watermark signal is public knowledge (a property that may e.g. also be true for copyright protection watermarks if they are employed as a deterrent), and that a detector may also be available to the public.

Unlike in copyright protection scenarios, no immediate adversarial relation needs to be considered since the marking itself is not directed against the interests of a particular individual or group.

Given the robustness of most digital audio watermarks particularly to digital/analog conversions, one of the prime uses of annotation watermarks is the association of an analog audio signal with its digital original; this can e.g. occur via centralized database records, limiting the payload requirement to a single unique key for such a database. The ability to reference the original audio track given possibly only a cropped or otherwise partial copy of a the whole audio material significantly eases record handling and can enable multimedia document management systems.

Regardless of whether the payload for the digital watermark is the annotation itself or a key into a database containing the actual referenced records, safeguards such as error-detecting or error-correcting codes must be employed to protect the integrity of the annotation. Conversely, some applications for annotation require that the integrity of the annotated signal be preserved. This can take the form of several possible sub-requirements; in a simple case the duplication and transfer of an annotation watermark without authorization to another carrier signal must be prohibited. A more elaborate requirement is that the semantic integrity of an annotation-marked signal must be preserved. This requirement can, to some extent be satisfied by the use of fragile watermarks, but – as noted in chapter 2 – this implicitly contradicts several robustness requirements.

The application scenarios in the next three sections for transaction tracking 3.4, active broadcast monitoring 3.5 and playlist generation 3.5.1 can be considered specific sub-scenarios of annotating audio signals via digital watermarks.

3.4 Transaction Tracking

To enable distribution tracking a necessary precondition is to identify the distributed copy of the audio track by means of a watermark. Therefore the watermark is comparable to a serial number identifying each of the sold watermarked copies as it is the case in the distribution of other digital products. One possibility is to embed the ID of the customer c_{id} as the audio watermark which does not identify each individual transaction uniquely but the customer as the potential head of an unauthorized redistribution chain. This corresponds to an application sub-scenario described in section 3.1.2 on page 40 and implies the same requirements for the payload size of the watermark. The watermark will be embedded into the audio file A^i to create the watermarked copy $A^i_{C_i}$ right after the authentification of the user C_i and before delivery in an audio-on-demand system (see figure 3.1 on page 47). In this scenario the generation of watermarks as well as a distribution medium that permits the efficient creation of distinct copies requires a considerable computational overhead in order to avoid wait states. This will become even more critical in case of multiple download requests. An effective system was presented by Arnold et al. [8, 6] in a way that the highest available bandwidth to be used without limitations imposed by constraints due to the speed of the used watermarking method.
Delivery of the same audio track to different customers or different music pieces to the same recipient allows the application of collusion (or *averaging*) attacks (see section 8.2.2 on page 158). In turn besides high robustness against signal processing the watermarks have designed to be collusion-secure [17] in order to make the watermarking technique applicable for transaction tracking.
The computational costs of the generation of distinct copies can be

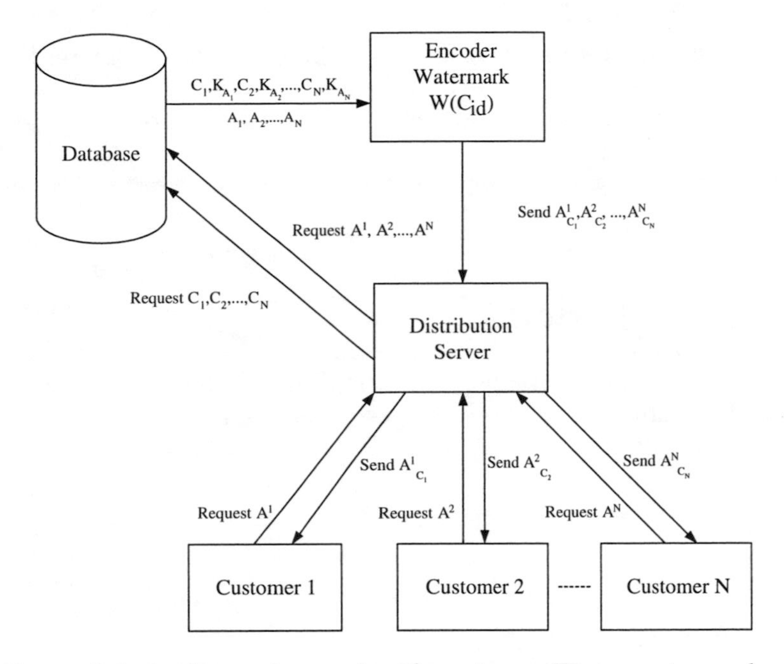

Figure 3.1: Audio on demand with customer IDs as watermarks

shifted from the content provider to the end user site if a playback device that contains a subsystem tied to a specific individual or customer is embedding the watermark immediately on playback. This scenario was e.g. used by the now-defunct DiVX pay-per-view digital video player scheme developed and owned by a partnership between Ziffren, Brittenham, Banca & Fischer, an entertainment law firm in Los Angeles, CA, and Circuit City.
The major drawback of marking on playback is that both the embedding mechanism and the requisite key material are present in the playback device (even download of ephemeral key material does not alter this situation) and hence under the control of a potential adversary. Moreover each playback device needs a separate embedding key since otherwise any single compromised playback device

would mean that the adversary can embed arbitrary customer IDs, eliminating the evidentiary value of the watermarks.
However, this implicit requirement for separate keys for each device imposes a severe computational burden in case a device watermark needs to be identified as for each suspect device, a test for the existence must be performed.

3.5 Monitoring

Switching to the air as the other distribution channel one of the most interesting applications is to embed a broadcast identification t_{id} into the audio stream. The watermark might carry an identifi-

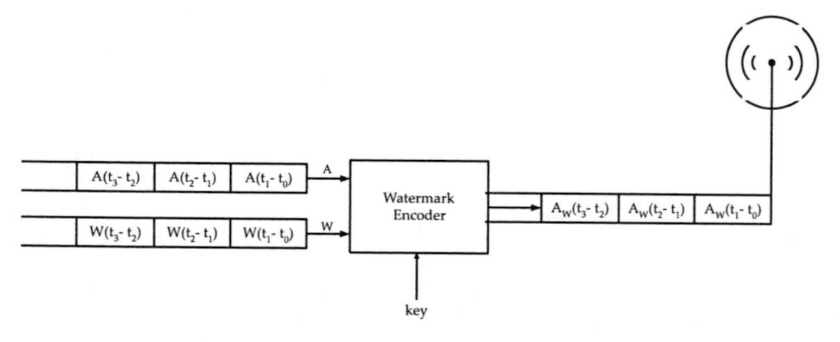

Figure 3.2: Embedding watermarks for transmission identification

cation ID and a timestamp specifiying the transmission time of the audio packet. In turn the content of the watermarks $W(t_j - t_i), t_i < t_j$ varies over the time (see figure 3.2). In the case of audio files being transmitted the watermark W can contain a timestamp t_{id} and an identification code like the International Standard Recording Code (ISRC). Embedding this type of information enables several different monitoring scenarios depending on the intention of the associated information broadcast along with the content and the time of monitoring right before or after the broadcast.

3.5.1 Automatic Playlist Generation for Rights Verification

Broadcasts of audio streams must be accompanied by appropriate compensation for the rights owners. While there exist societies offering centralized records keeping and compensation of individual rights owners in most jurisdictions (e.g. ASCAP for music in the United States), it is nonetheless a burdensome and error-prone work to create the requisite records on which creation was played, when it was played, and how many times.
The identification via a unique ID can be established by the detection of embedded annotation watermarks (see above) into each individual audio stream. Performing the monitoring on the originating site right before the broadcast implicitly determines the time when the song was played. This lowers the reporting burden on the performing entity and implies lowered robustness requirements due to a lower risk of distortions introduced by compression, transmission faults, and incomplete reconstruction. In such an application subscenario the bandwidth (typically time) required for a given digital watermark is of secondary interest.

3.5.2 Broadcast Monitoring of Audio Material

The automatic generation of playlists is also relevant for the reverse case, i.e. to verify the fact that a given creation (typically advertisement) has been broadcast according to a previously established contract. In addition monitoring of broadcast audio material is of interest not only for performers who want to ensure that they get the royalities from advertising firms but for copyright owners using the monitoring service to detect the illegal rebroadcast of their songs by pirate stations. Ideally, the copyrighted material is marked with a general copyright label as well as a label that identifies the customer who licensed the material in question. In a general setting, copyright owners license multiple different works. This results in one secret key being used to embed the copyright labels in all copyrighted works of that owner, such that the monitoring process can be applied uniformly to all copyrighted works.

However, this approach has a serious drawback – once the secret key is known, the copyright labels in all the works copyrighted by that owner can be removed easily. Unfortunately, there may be situations, where the owner of a copyright is forced to disclose his key to a third party, for instance in the course of an ongoing dispute over an alleged license violation.

One solution to this problem is to use unique keys for each copyrighted work. Consequently, the monitoring process must have efficient means of mapping a monitored work to the key that was used for watermark embedding which results in the so-called *identification problem*. This mapping can be established by using so-called *fingerprinting techniques* for the identification of the original track or the watermarking key respectively [86].

The threat of performing deliberate removal on the part of the entity broadcasting or performing the creations would be limited, as the lack of annotation watermarks would be sufficiently abnormal to warrant a manual inspection of the material presuming that the presence of annotation watermarks was distributed by the royalty-collecting entity. In turn, as in the previous case, the robustness requirements are derived mainly from the need to withstand standard processing chains encountered.

As opposed to the previous scenario of generating the playlists, the time required for the recovery of a digital watermark is critical to the overall efficiency of the monitoring scheme since, if each marking requires only a fraction of the duration of an audio track (five seconds for example), multiple signal sources can be monitored simultaneously by moving to a different source once a marking has been detected. Assuming that such a scan cycle does not last more than the average length of an audio track and a low unit value for individual royalty payments, this reduces the expenditures necessary for the monitoring equipment and bandwidth. Even more cursory monitoring would also be adequate since typically only gross or systematic underreporting is of actual interest.

3.6 Summary

This chapter discussed the use of watermarking in a variety of applications not limited to security related scenarios. The presentation showed that the watermarking technology has several advantages in comparison to other content protection techniques:

- The embedded watermarks are not audible which enables the embedding into the audio content itself.
- The watermarks are not separable from the original audio track.
- The watermarks undergo the same processing as the carrier signal.

These features makes the watermarking technology suitable in the following applications presented:

Copyright protection by watermarking can be established at different levels like identifying the origin of the audio track by means of the watermark and proving the ownership by the knowledge of the secret key for embedding and retrieval. The watermark technology is superior to traditional copyright notices since the copyright is transferred during copying and watermark are robust against format conversions.

Copy protection can be performed with watermarks in conjunction with specially equipped devices for playback and recording. The presence or absence of the watermarks enables the setup of playback and record control mechanisms in order to permit playback or to distinguish between first and subsequent copies of audio tracks.

Annotation of audio tracks by watermarks acts as a side channel reducing the effort in record handling and enabling multimedia document management systems. In contrast to other applications there is no adversial relation and the prime use is association of the audio signal to descriptive information.

Transaction Tracking uses watermarks to identify the watermarked copies sold in case of distribution.

Monitoring by means of the watermark can be performed in different scenarios before or after the audio material is broadcast. The watermarked audio stream can be monitored on the originating site before delivery for automatic playlist generation or after transmission to verify the broadcasting of the advertisement according to a contract or to detect illegal rebroadcast songs by copyright owners.

The suitability of the watermarking technology for each application has been discussed in detail regarding the requirements of the quality of the watermarked audio tracks, the robustness of the embedded watermarks, the playload size, the false positive rate, the security provided, and the speed for the embedding and retrieval processes.

CHAPTER
FOUR

Psychoacoustics

The preservation of the quality of watermarked audio tracks is one of the main requirements in all applications (see chapter 3). Therefore, the development of an effective digital audio watermarking technologies relies on detailed knowledge of the human auditory system. This chapter presents psychoacoustic facts and models exploited in the design of audio watermarking systems in order to embed an inaudible watermark.
Section 4.2 discusses the masking effect, the most important effect used for embedding an inaudible watermark signal into audio data. Furthermore the concept of critical bands is presented in section 4.3 describing the non-linearity behavior of our hearing system.

4.1 Introduction

The science of psychoacoustics describes acoustic from the perspective of the human auditory system. The abilities of the auditory system are not only investigated in the qualitative relation between sound and the corresponding impression, but also by quantitative relations between the stimuli presented and hearing sensations [109]. The relevant information is not only presented by the ability of the ear to hear frequencies in a band between 20 Hz and 20 kHz with the dynamic range of over 96 dB. The interaction of different frequencies

and the corresponding processing of the human auditory system is important to give a deeper understanding of the correlation between acoustical stimuli and hearing sensations.
The development of an exact model of the auditory system is a complex and to a certain extent subjective task. Physically sounds are easily described by the time-varying sound pressure $p(t)$. The processing of this sound pressure leads to a complex auditory sensation. The inputs to the human auditory system are the temporal variations in sound pressure. The processing in the auditory system leads to an output which contains information about the temporal and spectral characteristics of sound as well as the localization of the sound source.
The science of psychoacoustics tries to describe this information processing of the human auditory system. The most significant contributions in this field were made by Zwicker et. al [109]. As indicated above, the development of sophisticated models for the auditory system is based in large part on the use of extensive experimental data which of necessity implies a certain subjective aspect that can only be canceled through a sufficiently large data set.
Masking, pitch, critical bands, excitation and just-noticeable changes describe the active processing of the ear. Of these effects, masking plays the most important role in the lossy compression of digital audio data. Sophisticated models of masking in the frequency and time domain have been developed and applied to effective compression of audio data [49]. Data reduction by a factor of 12 (see table 4.1 on page 55) can be achieved without a significant loss of sound quality.

4.2 Masking Effects

Masking is an effect that occurs in everyday life. To enable a normal conversation the power of speech does not need to be very high. However, if emergency vehicles are passing by with loud sirens while we are talking on the street our conversation is nearly impossible. We normally have to wait until the emergency vehicle has passed or raise our voice to a greater loudness in order to continue the

Quality	Bandwidth	Mode	Bit rate	Reduction factor
telephone sound	2.5 kHz	mono	8kbps	96:1
better than short wave	4.5 kHz	mono	16kbps	48:1
better than AM radio	7.5 kHz	mono	32 kbps	24:1
similar to FM radio	11 kHz	stereo	56 ... 64 kbps	26. . . 24:1
near CD	15 kHz	stereo	96 kbps	16:1
CD	> 15 kHz	stereo	112. . . 128 kbps	14. . . 12:1

Table 4.1: Performance of the MPEG 1 Layer III compression algorithm

conversation. These effects also take place in the case of music, where louder instruments can mask out faint ones. This is a typical example of the so-called simultaneous masking.
Non simultaneous masking takes place when the masker and the test sound are not presented simultaneously but in close connection in time.[1] Two different situations are distinguished according to time relation of the the test sound and the masker:

- Premasking [2] is the effect in which the test sound is presented before the masker.
- Postmasking also called backward masking takes place when the test sound is presented after the masker disappeared.

Besides this total masking of sound there exists also a so-called partial masking effect which reduces the loudness of the test sound. Because this effect is not relevant in the case of audio watermarking it will not be considered here.
In order to measure the effect of masking quantitatively, the so-called masking threshold is determined. It is the sound pressure level of the test tone necessary to be just audible if a masker is switched on. The masking threshold is identical with the threshold in quiet if the frequencies of the masker and of the test tone are very different.

[1]Masker is the tone which masks out other sound. Test sound is the sound which will be masked.

[2]The expressions pre-stimulus and forward masking are also used.

The simultaneous and non-simultaneous masking phenomena are segregated by the temporal characteristics of the masker and test sounds. The next two sections explain this two different effects in masking. From the experimental facts presented below, psychoacoustic masking models have been developed and also integrated into perceptual audio encoders such as MPEG audio.

4.2.1 Simultaneous Masking

Steady-state conditions can be assumed if the test and masking sound have a duration longer than 200ms. The quantization error introduced by coding audio data can be seen as noise, therefore only the tonality of the masker, and not the masked noise, has to be considered.

Noise Masking Pure Tones Noise can be categorized by the physical term *spectral density*. Broad-band noise with constant spectral density results in a masking produced with a linear behavior (see figure 4.1). Below 500Hz the masking threshold is horizontal. Above

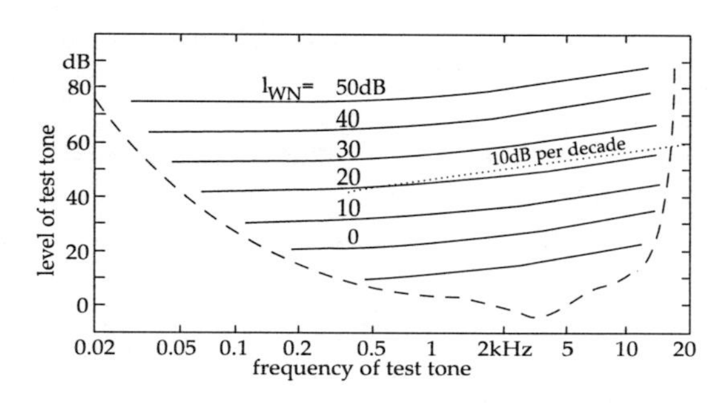

Figure 4.1: White noise masking pure tones

500Hz the masking threshold increases by a slope of about 10 dB per decade. Increasing the masking level of the masker by 10 dB shifts the masking threshold by the same 10 dB. The masking threshold is the same as the threshold in quiet for very low and high frequencies. Narrow-band noise possesses a bandwidth equal to or smaller than the critical bandwidth (100Hz for f < 500 Hz and $0.2f$ for f > 500 Hz, see section 4.3 on page 60). The masking threshold in this case depends on different parameters:

The center frequencies of the masker. The masking curve is broader for lower frequencies. Moreover the maximum of the masking curve decreases for higher frequencies (see figure 4.2).

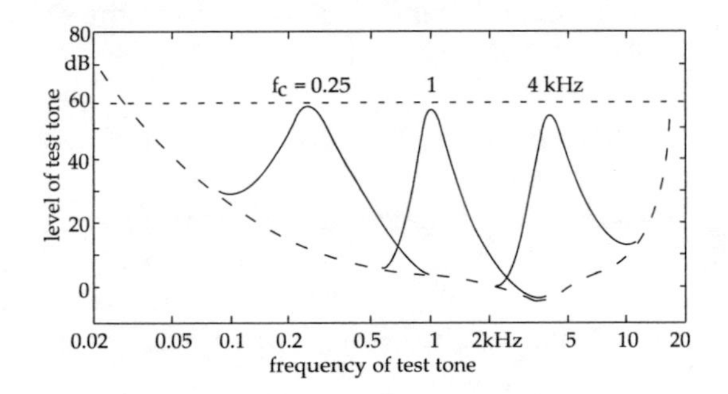

Figure 4.2: Narrow band noise masking pure tones

The level of noise masker presented. The slope of the rise from low to high frequencies before the maximum is reached is independent from the level of the masker. The masking threshold beyond the maximum decays lower for higher noise levels. Therefore the level dependence of the masking curve is a non-linear effect (see figure 4.3 on page 58).

Tones Masking Pure Tones As is the case for narrow-band noise the masking threshold for tones masking pure tones depends on the same parameters (see figure 4.4 on page 58):

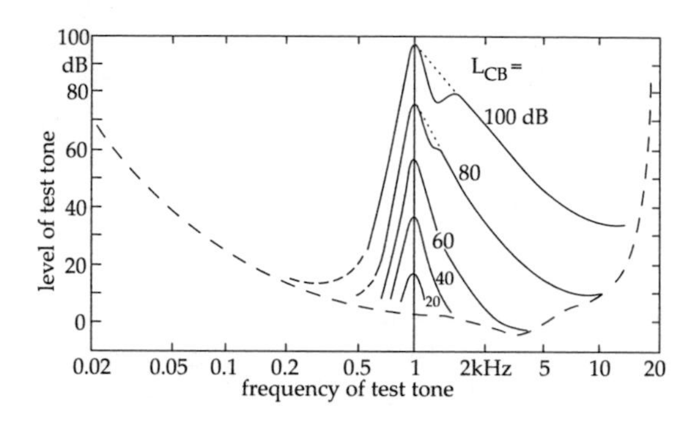

Figure 4.3: Level dependence of masking threshold

The level of the tone presented. In contrast to the narrow band noise masking the slope from lower frequencies to the frequency of the masker is more steep for higher masking level. The masking threshold beyond the maximum decays lower for higher tonal masker levels. Furthermore the maximum of the masked threshold is reduced with tonal maskers in comparison to narrow-band maskers.

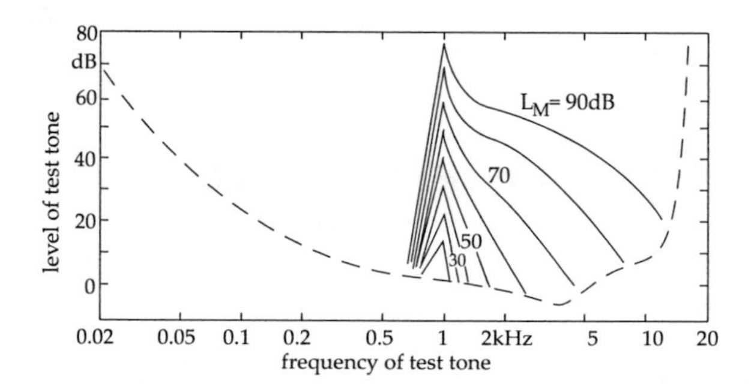

Figure 4.4: Masking threshold for tones masking pure tones

The frequency of the masker. Apart from frequencies lower than

500 Hz the masking curves of tonal masker with other frequencies can be approximately obtained by shifting the masking curves in figure 4.4 on page 58 such that the maximum of the masking threshold lies at the frequency of the tonal masker.

4.2.2 Temporal Masking Effects

Nearly every type of music has a strong temporal structure. Test and masking sound having a temporal characteristic produce so-called temporal masking effects. To quantitively measure the effects of temporal masking maskers of limited duration and short test-tone pulses have to be used. In order to measure the time relations between test tone and masker the test sound is shifted relative to the masker. According to the time shift Δt relative to the masker three different regions can be differentiated (see figure 4.5).

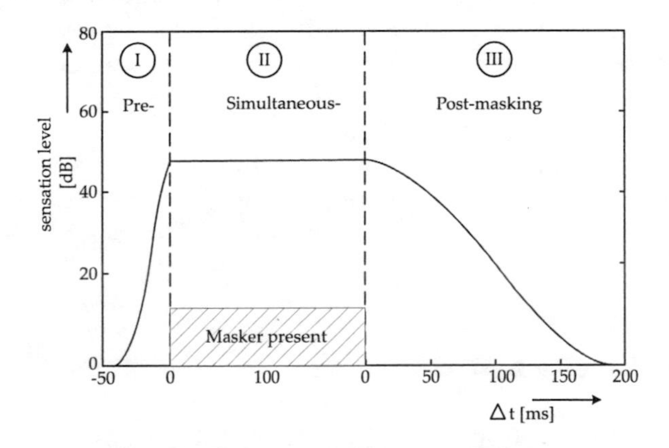

Figure 4.5: Regions with premasking, simultaneous masking and postmasking

The premasking effect happens before the masker is switched on in region I of figure 4.5. Premasking lasts about 20 ms in any condition. This means the threshold remains unchanged until Δt reaches

a negative value of 20 ms as shown in figure 4.5. After -20 ms $\leq \Delta t$ the threshold increases and reaches the level found in simultaneous masking just before the masker is switched on. The effect of premasking at first glance appears like listening into the future. Of course the information processing in our auditory system does not work instantaneously. The time needed to perceive the sound depends on the loudness of the presented sound. Therefore loud masker have a shorter set-up time then the faint test sound and will be perceived earlier.
The second region is the area of simultaneous masking. Threshold in quiet and masked thresholds depend on the duration of the test sound. This can be explained if we take into account that the hearing system integrates the sound intensity over a period of 200 ms. Therefore for durations of test sounds shorter than 200 ms the threshold in quiet and the masking thresholds increase because of the lower intensity due to the integration ability of the auditory system.
The third region in figure 4.5 describing postmasking corresponds to a decay of the masking effect after the masker is switched off. After a 5 ms delay the masking threshold decreases from the level it had in the simultaneous masking region.
At about 200 ms the level of the masking threshold reaches the threshold in quiet. Postmasking depends strongly on the duration of the masker. The decay of the masking threshold is much steeper for shorter maskers.

4.3 Critical Bands

Critical bands are an important concept in describing the auditory sensations. A corresponding construct, a so-called critical-band rate scale, was defined which is based on the fact the human auditory system analyzes a broad spectrum into different parts (see above). These parts are the so-called critical bands. Table 4.2 on page 61 was built by adding one critical band to the next in such a way that the upper limit of the lower critical band corresponds to the lower limit of the next higher critical band.

z / Bark	f_u / Hz	f_o / Hz	Δf / Hz	z / Bark	f_u / Hz	f_o / Hz	Δf / Hz
0	0	100	100	13	2000	2320	320
1	100	200	100	14	2320	2700	380
2	200	300	100	15	2700	3150	450
3	300	400	100	16	3150	3700	550
4	400	510	110	17	3700	4400	700
5	510	630	120	18	4400	5300	900
6	630	770	140	19	5300	6400	1100
7	770	920	150	20	6400	7700	1300
8	920	1080	160	21	7700	9500	1800
9	1080	1270	190	22	9500	12000	2500
10	1270	1480	210	23	12000	15500	3500
11	1480	1720	240	24	15500		
12	1720	2000	280				

Table 4.2: Critical-band rate z and corresponding frequencies

The critical bandwidth has a constant value of 100 Hz up to a center frequency of approximately 500 Hz. Above 500 Hz, a good approximation for the bandwidth $\Delta f/\mathrm{Hz}$ is 20% of the actual frequency. The following two analytic expressions are used to describe the dependence of critical band rate and critical bandwidth over the entire auditory frequency range:

$$z = 13\arctan\left(0.76\,\frac{f}{\mathrm{kHz}}\right) + 3.5\arctan\left(\frac{f}{7.5\mathrm{kHz}}\right)^2 \text{[Bark]} \quad (4.1)$$

$$\Delta f = 25 + 75\left[1 + 1.4\left(\frac{f}{\mathrm{kHz}}\right)^2\right]^{0.69} \text{[Hz]} \quad (4.2)$$

The critical bandwidth as a function of frequency shows the nonlinearity behavior of our hearing system (see equation 4.2). The critical band rate grows from 0 to 24 and has the unit Bark. The critical-band rate is related to several other scales that describe characteristics of the hearing system.

4.4 Summary

This chapter has described facts and models from the science of psychoacoustics, which provide the necessary background information for understanding the following chapters.
The most important effect is the masking effect which has been described in detail by a quantative measure, the so-called masking threshold. It describes the sound pressure level of the test tone (respectively the watermark signal) to be just audible if a masker is switched on. The dependence of the masking threshold on the frequency, tonality, and the level of the masker has been presented, which results in different masking curves. Furthermore the distinction between simultaneous and temporal masking effects has been discussed dependent on the time relation between the masker and the test sound.
Moreover the concept of critical bands has been presented which is necessary to understand audio watermarking algorithms presented in chapter 5 and the development of the audio watermarking method in chapter 6 of this dissertation.

CHAPTER

FIVE

Digital Audio Watermarking Methods

A number of approaches have been developed over the last few years to embed information into audio data, which will be presented in this chapter. The chapter starts with a presentation of the trivial Least Significant Bit (LSB)-Coding in section 5.1, a method which can be used for all types of media. Different features of the audio signal are used for embedding the watermark bits in other approaches. Two algorithms using the phase for information hiding are presented in section 5.2. The so-called echo hiding methods exploiting the temporal masking effects are described in section 5.3. Applying the watermarking technology to widely available compressed audio data by altering certain characteristics is presented in section 5.4. The last section presents spread-spectrum methods which enable the embedding of highly robust watermarks.
In the notation used, $\mathbf{c_o}[i], i = 1, \ldots, l(\mathbf{c_o})$[1] are the samples of the original signal in the time domain[2]. The range of sequence of numbers is according to the amplitude resolution of 8- or 16-bit $\mathbf{c_o}[i] \in \{0, 255\}$ or $\mathbf{c_o}[i] \in \{-32768, +32767\}\ \forall i$. An additional index of the carrier elements $\mathbf{c_o}_j$ denotes a subset of the audio signal. To the authors' best knowledge, all audio watermarking algorithms split the audio sig-

[1] $l(\mathbf{c_o})$ denotes the number of samples of track $\mathbf{c_o}$.

[2] If the audio data is sampled at a sampling rate $f_s = 44.1$kHz, one second corresponds to 44100 samples.

nal into different overlapping or non-overlapping blocks[3]. For this reason $\mathbf{c_o}_j[i]$ denotes the i-th sample in the j-th block (with length $l(\mathbf{c_o}_j)$). The individual blocks are used to embed part of one bit of information, one bit, a sequence of bits or the whole watermark denoted by $\mathbf{m}$. The length of the blocks are often determined by the usage of psychoacoustic models (see section 6.5 on page 105), the special transformation performed on the block, or the number of bits to be embedded.

5.1 LSB Coding

One of the first techniques investigated in the watermarking field as for virtually all media types is the so-called LSB encoding. It is based on the substitution of the least significant bit of the carrier signal with the bit pattern from the watermark noise. It uses no psychoacoustic model in order to shape the watermark. A natural approach in the case of audio data is to use least significant bit of the individual samples of the digitized audio stream having an amplitude resolution of for example 16 bit [4]. This *blind watermarking* method requires an exact synchronization of the marked audio data during the detection procedure. Besides having a high payload of 44.1 kBit/s, the low robustness makes this technique useless in real watermarking applications.

5.2 Embedding Watermarks into the Phase

Approaches which embed the watermark into the phase of the original signal do not use the temporal or spectral masking effects (see section 4.2 on page 54), but exploit the fact that the human auditory system has a low sensibility against relative phase changes [109].

[3]Usually with the same size.

[4]The usual amplitude resolution for audio files in CD format.

5.2.1 Phase Coding

The method presented by Bender et al. [14] splits the original audio stream into blocks and embed the whole watermark into the phase spectrum of the first block. The original signal $\mathbf{c_o}$ is split into $M = \left\lfloor \frac{l(\mathbf{c_o})}{N} \right\rfloor$ blocks $\mathbf{c_o}_j, 0 \leq j \leq M-1$ with $N := 2l(\mathbf{m})$ samples. The algorithm is structured as follows:

1. Each block of $\mathbf{c_o}$ is transformed in the Fourier domain $\mathbf{C_o}_j = \mathcal{F}\{\mathbf{c_o}_j\}, \forall j$. A matrix of the phases $\phi_{o_j}[\omega_k]$ and magnitudes $|\mathbf{A_o}_j[\omega_k]|, 0 \leq k \leq N/2 - 1$ is constructed.

2. The matrix with the differences in phase between the M neighbor blocks is computed:

$$\Delta\phi_{o_{j+1}}[\omega_k] = \phi_{o_{j+1}}[\omega_k] - \phi_{o_j}[\omega_k], \forall j, k \tag{5.1}$$

3. The watermark is encoded in the phase spectrum of the first block:

$$\phi_{w0}[\omega_k] = (-1)^{\mathbf{m}[k]+1}\frac{\pi}{2}, \text{ for } \mathbf{m}[k] \in \{0, 1\}, 0 \leq k \leq N/2 - 1 \tag{5.2}$$

4. In order to ensure the inaudibility of the phase changes between the consecutive blocks the phase differences in the blocks have to be adjusted:

$$\phi_{w_{j+1}}[\omega_k] = \phi_{w_j}[\omega_k] + \Delta\phi_{o_{j+1}}[\omega_k], \forall j, k \tag{5.3}$$

5. The original magnitudes $|\mathbf{A_o}|$ and the modified phase spectrum ϕ_w of the blocks are used to compute the marked signal in the time domain $\mathbf{c_w}_j = \mathcal{F}^{-1}\{\mathbf{C_w}_j\}\forall j$.

Before decoding the watermark, a preprocessing step is necessary in order to synchronize with the begin of the starting sequence. A necessary precondition in the decoding stage is the knowledge of the length[5] of the watermark $l(\mathbf{m})$

[5] The watermark length defines the number of samples via $N = 2l(\mathbf{m})$ used in the DFT.

1. Synchronization on to the first block $\mathbf{c_{w0}}$.
2. Transformation of the block $\mathbf{C_{w0}} = \mathcal{F}\{\mathbf{c_{w0}}\}$
3. Reading the bits of the watermark from the phase information of the first block $\phi_{w0}[\omega_k]$ for $0 \le k \le N/2 - 1$.

One disadvantage of the phase coding approach is the low payload that can be achieved. Only the first block is used for embedding the watermark. Moreover, the watermark is not distributed over the entire data set but implicitly localized and can thus be removed easily if cropping is acceptable.

5.2.2 Phase Modulation

Another form of embedding the watermark into the phase is by performing independent multiband phase modulation [60]. Inaudible phase modifications are exploited in this algorithm by controlled multiband phase alterations of the original signal. The original signal $\mathbf{c_o}$ is segmented into $0 \le m \le M - 1, M = \left\lceil \frac{l(\mathbf{c_o}) - N}{2N} \right\rceil$, blocks with N samples using overlapping windows. The window function is:

$$\text{win}[n] = \sin\left(\frac{\pi(2n+1)}{2N}\right), 0 \le n \le N-1 \tag{5.4}$$

Two adjacent blocks consist of the original and a watermarked block. The k-th watermarked block ($k = 2m$) carries the k-th sequence of the watermark. To ensure inaudibility by introducing only small changes in the envelope the phase modulation is performed by fulfilling the constraint given in equation (5.5)

$$\left|\frac{\Delta\phi(z)}{\Delta z}\right| < 30^o, \tag{5.5}$$

where $\phi[z]$ denotes the signal phase and zwhere $\phi[z]$ is the Bark scale according to equation (4.1). A slow phase change over time is achieved by using a long block size of $N = 2^{14}$. The algorithm proceeds as follows:

1. Each block of $\mathbf{c_o}$ to be watermarked is transformed in the Fourier domain $\mathbf{C_{o}}_k = \mathcal{F}\{\mathbf{c_o}_k\}$, $k = 2m, 1 \leq m \leq \left\lfloor \frac{M-1}{2} \right\rfloor$ yielding the Fourier coefficients $\mathbf{A_o}_k[f]$.

2. This step constructs the phase modulation function $\Phi_k(z)$. One integer Bark scale carries one information bit of the watermark. Each message bit is represented by a phase window function centered at the end of the corresponding Bark band and spans two Bark bands.

$$\phi(z) = \sin^2\left(\frac{\pi(z+1)}{2}\right), -1.0 \leq z < 1.0 \tag{5.6}$$

   The sign $\mathbf{a}_k[j] \in \{-1, 1\}$ of the phase window function is determined by the j-th message bit $\mathbf{m}_k[j] \in \{0, 1\}$ of the k-th sequence. The total phase modulation is obtained by the linear combination of the overlapped phase window functions:

$$\Phi_k(z) = \sum_{j=1}^{J} \mathbf{a}_k[j]\phi(z-j), 0.0 \leq z < J \tag{5.7}$$

3. Using the $\Phi_k(z)$, the bits are embedded into the phases in the k-th audio block by multiplying the Fourier coefficients with the phase modulation function:

$$\mathbf{A_w}_k[f] = \mathbf{A_o}_k[f] \times e^{i\Phi_k[f]} \tag{5.8}$$

   with f the frequency in Hz in contrast to the Bark scale z (see equation (4.1)).

4. The marked signal is computed by inverse transformation of the modified Fourier coefficients $\mathbf{A_w}_k$ of the individual blocks $\mathbf{c_w}_k = \mathcal{F}^{-1}\{\mathbf{C_w}_k\}, \forall k$. All blocks are windowed and overlap-added to create the watermarked signal.

The robustness of the modulated phase can be increased by using n_z Bark values carrying one message bit. Since one integer Bark carries one message bit the increment can be calculated by $\pm 15 \times n_z$ corresponding to equation (5.5) on page 66. For audio tracks sampled with sampling rate f_s and the number of critical bands N_B, the data rate is[6]

$$\text{\#bits per block} \times \text{\#blocks per second} = \frac{N_B}{n_z} \times \frac{f_s}{N} \tag{5.9}$$

Retrieving the watermark requires a synchronization procedure to perform a block alignment for every watermarked block by using the original signal. The watermark bits from the k-th audio block are recovered from the phase modulation $\widehat{\mathbf{\Phi}}_k$ of the possibly manipulated block. A matching of the individual segments of the modulated phase to the encoded bits would be possible if the phase modulation is not distorted by manipulations.
In contrast the retrieved $\widehat{\mathbf{\Phi}}_k$ is a noisy version of the modulated phase $\mathbf{\Phi}_k$ preventing an easy decoding of the k-th sequence of bits from the k-th audio block. Nevertheless according to equation (5.7) on page 67 the modulated phase $\mathbf{\Phi}_k$ can be viewed as a sequence of state transitions of the four possible transitions $(0 \rightarrow 0, 0 \rightarrow 1, 1 \rightarrow 0, 1 \rightarrow 1)$.
Besides decoding each bit individually, this enables the modeling of $\widehat{\mathbf{\Phi}}_k$ as a hidden Markov model and a determination of the single best concatened sequence of those possible transitions. The possible transitions $\mathbf{p}^{ij}(f) = \{p_t^{ij}(f)\}_{t=1}^{T}, 0 \leq i, j \leq 1$ can be calculated in advance for the frequency range (in Hz) used for embedding the individual bit. T is determined by the number of bits embedded and f covers the frequency range in Hz for the t-th bit. The detection procedure is structured as follows:

1. Calculate phase modulation function $\widehat{\mathbf{\Phi}}_k$ by applying the window function in equation (5.4) on page 66 and performing the Fourier transformation $\mathbf{C_{w}}_k = \mathcal{F}\{\mathbf{c_w}_k\}$ of the k-th block.

2. Formulate the $\widehat{\mathbf{\Phi}}_k$ as an observation sequence $\mathbf{o}(f) = \{o_t(f)\}_{t=1}^{T}$ where f covers the frequency range in Hz for the t-th bit.

[6] $N_B = 24$ for $f_s = 44.1kHz$ according to table (4.2) on page 61.

3. Calculate the weight factor sequence $\boldsymbol{\beta}(f) = \{\beta_t(f)\}_{t=1}^{T}$ with

$$\beta_t[f] = \min\left(\left|\mathbf{A_w}_t[f]\right|^2, \left|\mathbf{A_o}_t[f]\right|^2\right), f = 0,\ldots,K-1 \quad (5.10)$$

$$\sum_f \beta_t[f] = 1 \quad (5.11)$$

   The weight factors of the t-th observation sequence are determined by the smaller spectral energy of the original or watermarked signal. This is based on the assumption that smaller spectrum components and their corresponding phase information are more likely to be distorted by some kind of non-linear processing like MPEG [49] encoding.

4. Calculate the cost function

$$c_t^{ij} = \frac{1}{K}\sum_{f=0}^{K-1}\left|(p_t^{ij}[f] - o_t[f])\beta_t[f]\right|, 0 \leq i,j \leq 1, 1 \leq t \leq T \quad (5.12)$$

5. Perform the Viterbi [85] search algorithm with the calculated cost function in order to find the best sequence of possible state transitions, which in turn yields the k-th sequence of bits.

Both phase embedding approaches use the psychoacoustic features of the human auditory system with regard to the just noticeable phase changes. They exploit the inaudibility of phase changes if the time envelope of the original signal is approximately preserved. Given the requirement for phase alterations, both embedding and retrieval of the digital watermarks are performed blockwise in the Fourier domain. While the phase coding method described in section 5.2.1 on page 65 is embedding the watermark in the phases of the first block, the phase modulation algorithm performs a long-term multiband phase modulation. Both algorithms are *non-blind watermarking* methods, since they require the original signal during the watermark retrieval, which of course limits their applicability.

In a copyright protection application the owner is authenticated by the original signal since no secret parameter is included in the algorithm. However, protocols must be designed carefully to avoid vulnerability to attacks described in section 8.2.4 on page 167.

5.3 Echo hiding

A variety of watermarking algorithms [37, 75, 58, 30, 106] are based on so-called *echo hiding* methods. Echo hiding algorithms embed watermarks into a signal $c_o(t)$ by adding echos $c_o(t - \Delta t)$ to produce a marked signal $c_w(t)$:

$$c_w(t) = c_o(t) + \alpha c_o(t - \Delta t) \tag{5.13}$$

Equation (5.13) contains two parameters which can be changed in order to provide inaudibility of the watermark and to embed the bits into the audio signal. The change of the delay time Δt is used to encode the bits of the watermark, whereas both parameters α and Δt have to be adjusted to ensure inaudibility of the embedded echo. In general equation (5.13) can be written as

$$c_w(t) = \sum_{k=0}^{N} \alpha_k c_o(t - \Delta t_k) \tag{5.14}$$

where $c_o(t)$ is the original signal with $\alpha_0 = 1, \Delta t_0 = 0$ and N the number of different echo signals embedded. Using the response function

$$h(t) = \sum_{k=0}^{N} \alpha_k \delta(t - \Delta t_k) \tag{5.15}$$

this can be written in short form as a convolution of these echos with the original signal

$$c_w(t) = c_o(t) * h(t) \tag{5.16}$$

In turn the marked signal $c_w(t)$ can be expressed in the frequency domain as

$$C_w(\omega) = C_o(\omega) H(\omega) \tag{5.17}$$

where $C_o(\omega)$ and $H(\omega)$ are the Fourier transformations of the signals $c_o(t)$ and $h(t)$ respectively. During the detection step the calculation of $h(t)$ is necessary in order to determine the individual echos with corresponding delay times Δt_k encoding the bits $k = 1, \ldots, N$. According to equation (5.17) the signal can be deconvolved by dividing $C_w(\omega)$ by $C_o(\omega)$ in the frequency domain and calculating the inverse Fourier transformation. Performing this operation requires an *a priori* knowledge of the original signal $C_o(\omega)$ which is not practical in the case of watermarking. On this account the detection method uses the so-called *homomorphic deconvolution* technique in order to separate the signal and the echos.
The basic idea behind homomorphic deconvolution is to apply a logarithmic function to convert the product (5.17) into a sum. Using the definition of the *complex Cepstrum* as the inverse Fourier transformation of the log-normalized Fourier transform of the watermarked signal the transformed signal can be written as

$$C_w(q) = \mathcal{F}^{-1}\{\log|C_o(\omega)H(\omega)|\} \quad (5.18)$$

$$= \mathcal{F}^{-1}\{\log|C_o(\omega)|\} + \mathcal{F}^{-1}\{\log|H(\omega)|\} \quad (5.19)$$

$$= C_o(q) + H(q) \quad (5.20)$$

as a function of the time or *quefrency*[7] domain. According to equation (5.18) the original signal $C_o(q)$ and the embedded echos $H(q)$ are clearly separated on the quefrency axis q. Using this deconvolution technique in the detection of the watermark bits, an algorithm adding two different echos for embedding '0'- and '1'-bit can be constructed. The original signal $\mathbf{c_o}$ is split into $M = \left\lfloor \frac{l(\mathbf{c_o})}{N} \right\rfloor$ blocks $\mathbf{c_o}_j, 0 \leq j \leq M-1$ with N samples. Each block carries one bit of the watermark. The embedding procedure proceeds as follows:

1. For each block $\mathbf{c_o}_j$ of the original signal the echo signal for the 0- and 1-bit are constructed with the corresponding delay time and attenuation factors α_0 and α_1.

$$w_k(t) = \alpha_k c_o(t - \Delta t_k), \text{ for } k = 0, 1 \quad (5.21)$$

[7] q is the quefrency and has the same units as time.

2. Two complementary modulation signals $m_k(t), k = 0,1$ for the 0- and 1-bit are generated:

$$m_0(t) = \sum_{j=0}^{M-1} (1 - b_j)\, rect_j(t), \quad m_1(t) = \sum_{j=0}^{M-1} b_j\, rect_j(t) \tag{5.22}$$

$$m_0(t) + m_1(t) = 1\, \forall t \quad rect_j(t) = \begin{cases} 1 & \text{for} \quad t_j \leq t < t_{j+1} \\ 0 & \text{otherwise} \end{cases} \tag{5.23}$$

   and

$$b_j = \mathbf{m}[j \bmod l(\mathbf{m})] \tag{5.24}$$

   The modulation signals are used to construct the echo signals according to the bits of the watermark.

3. After multiplying the echo signals $w_k(t)$ with the modulation signals $m_k(t)$, the marked audio stream is generated by the addition of the computed signals to the original one

$$c_w(t) = c_o(t) + m_0(t)w_0(t) + m_1(t)w_1(t) \tag{5.25}$$

Retrieving the watermark requires a synchronization procedure to perform an alignment to the watermarked blocks. After the synchronization is established the detection algorithm is structured as follows:

1. Transformation of the sequence in the Cepstrum domain $\mathbf{C_w} = \mathcal{F}^{-1}\{\log(|\mathcal{F}\{\mathbf{c_w}\}|)\}$.

2. Auto-correlation of $\mathbf{C_w}$ in the Cepstrum domain.

3. Measurement of the delay time δt via the peaks of the auto-correlation of $\mathbf{C_w}$.

4. Determination of the embedded bit by comparison of δt with $\Delta t_k, k = 0$ or 1.

Out of the masking effects presented in section 4.2.2 on page 59, the echo hiding approach uses the postmasking effect in order to control the inaudibility of the embedded watermark. The delay times Δt_k and attenuation factors $\alpha_k, k = 0, 1$ have to be adjusted in the embedding process according to the perception threshold of the human auditory system (see figure 4.5 on page 59) to ensure the inaudibility of the echoes. It is a *blind watermarking* method, which modulates the bits as echo signals embedded in individual blocks of the audio stream.
In contrast to the majority of audio watermarking algorithms the embedding and the detection are performed in two different domains, the time and Cepstrum domain, respectively. A disadvantage is the complexity of this method due to the number of transformations (see equation (5.18)) which have to be computed for detection, which is performed in the Cepstrum domain. Furthermore one major drawback of this approach is the vulnerability to malicious attacks since the information can be detected by anyone without using a secret key. An attacker can exploit this knowledge if he/she knows the underlying algorithm to apply a removal attack (see section 8.2.2 on page 155), as demonstrated by Petitcolas et al. in [78]. A possible countermeasure presented by Ko et al. [58] against the easy determination of delay time δt in the detection procedure (see above) is the spreading of the echo over the time axis. This is accomplished by substituting the Dirac delta function in the response function (5.15) by a pseudo noise (PN) sequence. Instead of calculating the auto-correlation in the Cepstrum domain, despreading of the echo is performed by cross-correlation of the cepstral signal in equation (5.18) with the PN sequence generated from a secret key.

5.4 Watermarking of Compressed Audio Data

Since a lot of audio tracks already published in the Internet are compressed versions of the original one, an obvious approach is to use the compressed audio material for embedding the watermarks. The two approaches presented below can be distinguished by the domain where the watermark is embedded and the domain where the detection of the watermark information is performed.

5.4.1 Watermarking the compressed bit-stream

Several approaches exist to embed the watermark directly into the already compressed audio bit-stream (see [70, 62, 32, 23]). As a result time consuming decoding, watermarking embedding, and re-encoding in the case of PCM watermarking techniques are not necessary in order to embed the watermark. In this case the retrieval process does not involve a decoding procedure, which results in an additional decrease in watermark retrieval speed. Nevertheless, the starting point for professionally created audio material is always the PCM format.[8] These approaches change the contents of the MPEG frame (see figure 5.1) directly.

Header 12 Bit Sync signal 20 Bit System information	Error Correction Code 16 Bit optional	Code for number of scale factors 2 Bit	Bit assignment 4,3,2 bit for lower, middle, upper band	Scale factors 6 bit	Sample values 2...15 Bit	Optional data

Figure 5.1: MPEG-Frame Layer III

The scaling factor can be viewed as a logarithmic gain factor for the sample values in order to retrieve the original samples in PCM format. The embedding of the watermark is done by changing the scaling factors of different frames according to a special pattern derived from a secret key. A problem of this method is that some audio streams carry only a few scaling factors per frame. Therefore, the space for embedding a watermark is reduced. This leads to the problem that multiple watermarks cannot be embedded using this approach, because altering scale factors already used for embedding the first watermark destroys the quality of the audio data.
A second approach in the variation of the MPEG frame [70] tries to alter the sample values instead of the scaling factors. Embedding

[8]Besides live recordings occasionally made on MiniDisc or sound tracks on MiniDV video camcorders.

multiple watermarks is also critical in this case. The additional requirement of using the original track as input for the retrieval process further limits the applicability of this approach.

Besides methods working on MP3 bit-streams methods like the one presented by Cheng et al. [23] are embedding watermarks into the AAC [19, 50] compressed bitstream by direct modification of the quantized coefficients. The watermark bits are embedded by performing a spread spectrum modulation (see section 5.5 on page 77) of the quantized coefficients. The individual bits are retrieved by a linear correlation of the PN sequence used during the embedding and the quantized coefficients of the watermarked bit-stream. The coefficients to be modified are selected by applying a heuristic, which uses only non-zero coefficients in a predefined frequency range. The amount of distortion applied is fixed and set to the quantization step size of 1.

Methods of directly watermarking the compressed bit-stream have the disadvantage that they do not make use of a psychoacoustic model. Both embedding and detection are performed directly on the compressed bit-stream, where the audio stream is processed in frames according to the formatting of the bit-stream in the specific compression algorithm. Additional information is not necessary if the audio data is synchronized. The main advantage is the low computational cost. Furthermore these methods provide implicit robustness against their specific compression format due to embedding of the watermark in the already compressed bit-stream. The main disadvantage of these methods is the missing psychoacoustic component in comparison to the uncompressed audio signal. The influence on the audio quality of the original track by altering scaling factors, sample data, or the quantized coefficients can only be estimated. Moreover, the decoding of the compressed bit-stream and a new compression with slightly shifted the audio stream may lead to a synchronization problem because of the new scaling factors, sample data, and quantization coefficients of the compressed stream. Furthermore the complexity advantage is lost if the watermarked audio tracks have to be transcoded to another compression format.

5.4.2 Integrating Watermark Embedding into Compression Encoder

Besides directly watermarking the bit-stream, other methods extract the information in the compressed bit-stream from the quantization of the audio samples [71]. This enables the estimation of the masking threshold to shape the watermark noise below this threshold in order to ensure inaudibility. Integrating the watermark and compression encoder has two advantages: The quality during the watermarking can be controlled in contrast to the methods described above and the speed of embedding is improved in comparison to two separate processes of watermarking and compression. The building blocks consist of parts of the PCM watermark embedder, and the compression decoder and encoder (see figure 5.2).

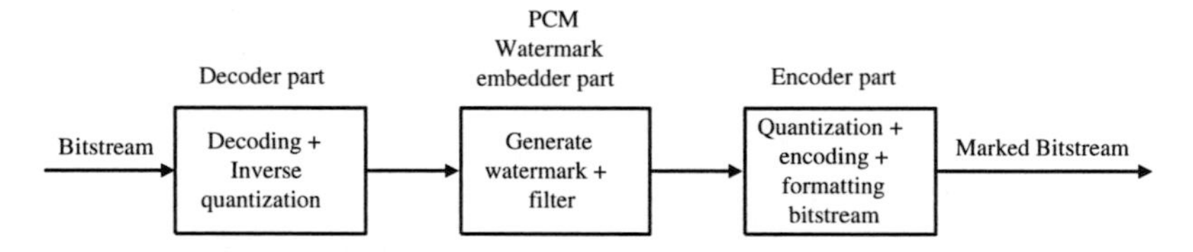

Figure 5.2: Integration of watermark embedding into compression encoder

Part of the bit-stream decoder is used in order to read the scaling factors and decode the bit-stream and perform the inverse quantization of compressed samples. The information about the quantization enables the calculation of the masking threshold. The masking threshold controls the multiplication factors used to multiply the spectral lines of the constructed watermark – as is typical for a perceptual watermark encoder – applying the masking effects.
The watermark generation can be the same as for the PCM watermark embedder. After weighting the spectrum of the watermark noise the result is added to the original spectral lines. The extracted scaling factors from the original frame are used in order to quantize

the marked audio data again and format the bit-stream. The final output is the marked bit-stream (see figure 5.3).

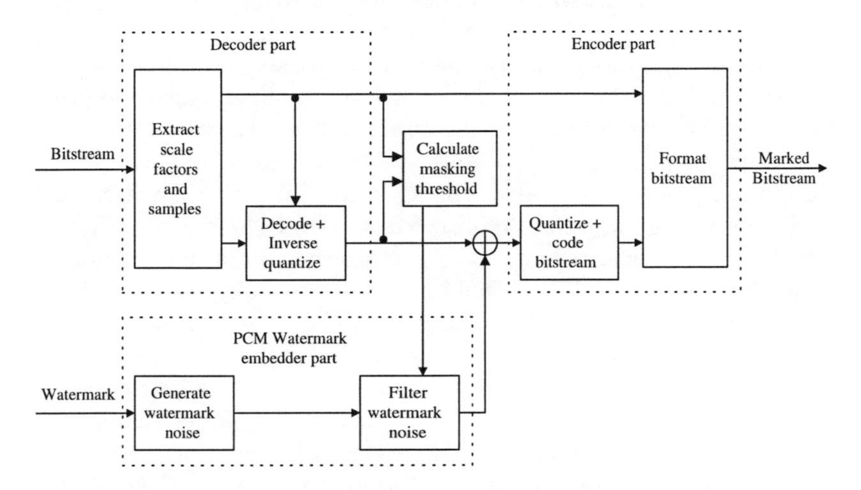

Figure 5.3: Components of bit-stream watermarker

This method makes implicit use of the psychoacoustic model by approximating the perceptual information contained in the MPEG frames. Detection can be performed on the compressed and uncompressed audio data. It is a *blind watermarking* method, which distributes the bits over different MPEG frames. Due to the usage of parts of the compression encoder and decoder, such a mechanism is tied to the special compression scheme used. For each newly developed compression algorithm, a new integration of the watermarking embedding procedure becomes necessary.

5.5 Spread-Spectrum Audio Watermarking

Spread-spectrum methods, originally conceived for masking the origin of radio transmissions and enhancing resilience against jamming,

are often used in the transmission of digital information. Since the requirements of suppressing of jamming during transmission, hiding a signal against unintended listener, and assuring information privacy are very similar to the one in watermarking applications these are probably the most widely used techniques in the development of watermarking algorithms. From the spread spectrum viewpoint the original audio signal can be considered as a jammer interfering with the signal carrying the watermark information.
The spread spectrum modulation is a special form of watermark modulation. The modulation is performed on $\mathbf{C_o}$ which is the transformed block of samples $\mathbf{c_o}$. The transformation is used to model the audio signal by orthonormal base functions spanning the signal space. If the identity transformation is used the signal is represented by the block of PCM samples itself. In case of the Fourier transformation, the trigonometric functions are used as basis functions and the transformed block consists of the Fourier coefficients represented by vector $\mathbf{C_o}$. Each bit $k \in \{0, 1\}$ is modeled by a pseudo-noise $\mathbf{pn}_k$ vector consisting of two equiprobable elements $\{-1, +1\}$ generated by means of the secret key. Consequently the expectation value of the pseudo noise sequence is $E\{\mathbf{pn}_k\} = 0$. Usually the pseudo noise sequences for the two bits are inverted $\mathbf{pn}_0 = -\mathbf{pn}_1 = \mathbf{pn}$. The original signal $\mathbf{c_o}$ is split into $M = \left\lfloor \frac{l(\mathbf{c_o})}{N} \right\rfloor$ blocks $\mathbf{c_o}_j, 0 \leq j \leq M-1$ with N samples. To simplify the discussion one block ($\mathbf{c_o} := \mathbf{c_o}_j$) carrying one bit of the watermark[9] is considered. The embedding algorithm is structured as follows:

1. The block $\mathbf{c_o}$ is transformed with the orthogonal transform $\mathcal{T}$ in the corresponding domain $\mathbf{C_o}$.

$$\mathbf{C_o} = \mathcal{T}(\mathbf{c_o}) \tag{5.26}$$

2. The PN sequence $\mathbf{pn}_k$ is weighted with α to adjust between quality and robustness.

$$\mathbf{W} = \alpha \mathbf{pn}_k \tag{5.27}$$

[9]The pattern can also be distributed over several blocks.

3. The modulated and weighted watermark signal is added to the cover signal in the transformed domain.

$$\mathbf{C_w} = \mathbf{C_o} + \mathbf{W} \tag{5.28}$$

4. The watermarked signal is transformed back into the time domain.

$$\mathbf{c_w} = \mathcal{T}^{-1}(\mathbf{C_w}) \tag{5.29}$$

During the detection step the same vector $\mathbf{pn}_k, k = 0, 1$ has to be generated via the secret key. A comparator function is used in order to decide about the presence of the embedded vector **pn**. This requires a perfect synchronization to the embedding block of samples. The detection procedure proceeds as follows:

1. Synchronization with the beginning of the embedding block $\mathbf{c_w}$.

2. Transformation of $\mathbf{c_w}$ into embedding domain $\mathbf{C_w} = \mathcal{T}(\mathbf{c_w})$.

3. Correlate $\mathbf{C_w}$ with $\mathbf{pn}_k, k = 0, 1$ by applying the comparator function C_τ

$$C_\tau(\mathbf{C_w}, \mathbf{pn}) = C_\tau(\mathbf{C_o}, \mathbf{pn}) + C_\tau(\alpha\mathbf{pn}, \mathbf{pn}) \tag{5.30}$$

4. Detection of the transmitted bit is usually made on the sign of the comparator function

$$sign(C_\tau(\mathbf{C_w}, \mathbf{pn})) \begin{cases} > 0, & \text{for } \mathbf{pn}_0 \\ < 0, & \text{for } \mathbf{pn}_1 \end{cases} \tag{5.31}$$

One of the widely used comparator functions C_τ is the linear correlation:

$$C_\tau(\mathbf{x}, \mathbf{y}) = \langle \mathbf{x}, \mathbf{y} \rangle = \frac{1}{N} \sum_{i=1}^{N} \mathbf{x}[i]\mathbf{y}[i] \tag{5.32}$$

with the signal vectors $\mathbf{x}$ and $\mathbf{y}$. The result of the correlation $C_\tau(\mathbf{C_w}, \mathbf{pn})$ consists of the two contributions $C_\tau(\mathbf{C_o}, \mathbf{pn})$ and $C_\tau(\alpha\mathbf{pn}, \mathbf{pn})$ as can be seen from equation (5.30). The second term accumulates the contribution of the pseudo noise sequence[10] embedded in the different base functions, whereas the first term represents the correlation or the interference of the carrier signal respectively and pseudo noise sequence.

Several audio watermarking algorithms use different embedding domains and representations of the transformed signal vector $\mathbf{C_o}$. Furthermore, the psychoacoustic parameters have to correspond to the specific embedding domain in order to perform the psychoacoustic weighting step.

One of the first algorithms which used the masking properties human auditory system by Tewfik et al. [18, 97] operates in the Fourier domain. The psychoacoustic weighting is performed by shaping the Fourier coefficients of the PN sequence according to the masking threshold calculated by the psychoacoustic model presented in section 6.5.1 on page 105. Furthermore this algorithm approximates the temporal masking behavior (see figure 4.5 on page 59) by using the envelope of the signal for the increase and a decaying exponential for the decrease of the signal, respectively. Another algorithm presented by Haitsma et al. [38] also embeds the watermark in the Fourier domain by altering the frequency magnitudes. The algorithm presented by Kirovski et al. [57] uses the modulated complex lapped transform (MCLT)[11] and modifies the magnitude of the MCLT coefficients in the dB scale rather in linear scale. They use a psychoacoustic model [56], which quantifies the audibility of the MCLT magnitude coefficient. The algorithm proposed by Bassia et al. [12] works in the time domain by altering the amplitudes of the samples. The shaping of the watermark is applied by performing an low-pass filtering of the PN sequence.

Spread-spectrum methods are widely used techniques for different types of media given their high robustness against signal manipula-

[10]This is often denoted as despreading the sequence.

[11]The MCLT transformation is a 2x oversampled DFT filter bank.

tions. By using a secret key to generate the pseudo noise sequence **pn**, these algorithms do not need the original audio signal in order to detect the embedded bits and are therefore *blind watermarking* methods provided that the synchronization requirement is met. The main disadvantage of spread-spectrum methods is the vulnerability against the de-synchronization attacks.

5.6 Summary

Several algorithms for embedding watermarks into audio data were presented in this chapter. The techniques described range from the simple LSB method to sophisticated spread-spectrum methods. Features used for embedding the watermark bits are the amplitude in the time domain, magnitude, frequency and phase in the Fourier domain and characteristics of the compressed audio stream.
The presentation of each method included a categorization with regard to the terminology introduced in chapter 2 for the different possible watermarking system. This was supported by the discussion of the properties of the system in terms of payload, security and robustness of the embedded watermark. Furthermore the psychoacoustic features exploited in the watermark embedder for hiding an inaudible watermark were detailed for each technique. This discussion enables the judgment of the suitability of each method for different applications presented in chapter 3. Moreover, the investigation of advantages and disadvantages of each method presented influenced the design of the audio watermarking method which will be presented in the following chapter.

CHAPTER

SIX

The A©WA Algorithm

A first step in the development of a reliable audio watermarking algorithm is to detail the various features of an optimal method, which will be defined in section 6.1. Considering the wide range of partially conflicting requirements, section 6.2 discusses their priorities and presents a ranking useful during the design of the audio watermarking algorithm.

This chapter describes the development of an effective digital watermarking algorithm from requirements through algorithmic design criteria, and several refinement steps. The basic so-called A©WA (Audio ©opyright protection by WAtermarking) algorithm is the subject of section 6.3. The theoretical background of the algorithm is presented in detail for embedding a single bit into the audio stream. While the embedding and detection of one bit of information is in principle sufficient in applications like copyright protection, embedding of several bits is necessary for scenarios where information embedded has to be retrieved. Furthermore the embedding of multiple watermarks in the same audio track can serve different purposes like copyright owner and customer identification. The necessary extensions of the basic method to embed more than one watermark simultaneously in the audio file with n bits are described in section 6.4.

Preserving the quality of the watermarked audio tracks is the most important requirement of an audio watermarking method. Conse-

quently the shaping of watermark with respect to the psychoacoustic facts by using appropriate models is of vital importance for the applicability of the proposed method. The psychoacoustic model used and the application to the watermarking problem are presented in sections 6.5.1 and 6.5.2.
Besides quality, robustness of the embedded watermarks represents the second most important requirement. Several improvements of the basic algorithm can be made both on the embedding and detection sides to improve the performance of the robustness and are the subject of section 6.6. Finally, section 6.7 summarizes this chapter.

6.1 Requirements of an Audio Watermarking Algorithm

According to the intended application of watermarks in audio data the algorithm as well as the watermark itself has to fulfill a set of requirements [18]. The international federation of phonographic industry (IFPI) has specified the desired features of an optimal audio watermarking method. These requirements can be elaborated and further subdivided into different categories describing the properties of the algorithm and the watermark

- Signal processing properties
- Security properties
- Application specific requirements

6.1.1 Quality

One of the main requirements especially in the audio field is to assure the *quality* of the watermarked copy. The is equivalent to the inaudibility of the watermark. An audible watermark would essentially render the audio track useless. For this reason the usage of psychoacoustic models (see section 6.5 on page 105) in order to ensure quality is one of the main cornerstones in developing a high quality watermarking method.

6.1.2 Robustness

A perfect audio watermarking method as designated by the requirements has to be *robust* against any kind of attack or manipulation which renders the watermark undetectable. At present no such system exists and it is not at all clear whether it will be possible to build a perfect watermarking system in the future. As a result one invariably tries to find a compromise between the competing and at least to a certain extent mutually exclusive requirements of quality, robustness, and payload size. The application can also determine the domain for embedding the watermark. As an example, the robustness against lossy compression is one of the main requirements concerning robustness for audio watermarking in the consumer area. An integration of the watermark embedding system into the compression encoder implicitly guarantees a higher resilience of the watermark (see section 5.4.2 on page 76) against this specific type of modification.
The catalogue of possible manipulations is depending on the application and the requirements contains, but is not limited to, the following signal manipulations [97]:

- Addition of multiplicative and additive noise.
- Low- and high-pass filtering.
- Lossy compression, for example MPEG 1 Layer I, II, III
- Noise reduction applying different types of algorithms.
- Changing the amplitude resolution of the audio samples.
- Altering the sampling rate of the audio track.
- Scaling the time axes.

- Cropping or insertion of samples are manipulations which can occur during the processing of audio tracks.
- Digital/Analog (D/A)- and A/D-Conversion.
- Distortions in the frequency domain.
- Overmarking (the additional embedding of a second watermark with the same or an other algorithm). This aims at destroying the already embedded watermark or embedding an own watermark to be able to claim ownership.
- Collusion attacks are applicable to areas which require the simultaneous embedding of several watermarks as it is the case in customer identification scenarios (see section 3.4 on page 46).
- Statistical attacks aim at the destruction of the watermark by means of averaging over several watermarked copies of the same original.

The different manipulations of watermarks (i.e. as opposed to semantic attacks such as overwatermarking and collusion) can be separated into two groups:

- Attacks which try to destroy the watermark.
- Synchronization attacks which try to misalign the embedded watermark and the corresponding detector in order to prevent the detection.

6.1.3 Security

Most of the watermarking applications have to ensure the security of the embedded watermark. Applications like the embedding of meta data into the audio stream on the other hand do not require a high security standard since the intention is to deliver information as an additional benefit to the user. The following aspects must be taken into consideration for each application scenario:

Secrecy of the key and the watermark The watermarking procedure should rely on a key to ensure security, not on the algorithm's secrecy [55]. The key is the secret parameter unknown to anyone but the authorized individuals embedding and detecting the watermark. Furthermore, the watermark should be inaudible and statistically undetectable to prevent attacks relying on the statistical features of the watermarking process or the watermark. Besides the fact that an audible watermark would reduce the quality, it would implicitly localize the embedded information. In order to improve the security, an additional encryption of the watermark can be performed before the embedding.

Counterfeiting of the watermark Besides the destruction of the watermark by signal processing or detecting the watermark one can try to construct a fraudulent watermark. Consequently the algorithm should prevent reading a watermark which is not actually embedded from the data set.

Mathematical formulation of the algorithm No watermark can be detected with 100% accuracy. In order to measure the security of the detection, false positive and false negative error probabilities are used as a metric. Accordingly the watermarking procedure should have a mathematical formulation to permit the validation as well as the adjustment of the level of security according to the intended application.

6.1.4 Application Requirements

Besides the general requirements as outlined above, there are additional more application-oriented demands concerning the integration of the watermarking algorithm into other applications:

Speed of watermark encoder and decoder To enable the identification of consumer buying the audio tracks in a commercial scenario it is necessary to embed a watermark for identification purposes (see section 3.4 on page 46) into the data immediately before delivering the content. Under these circumstances the speed of the encoder plays an important role in order to speed up the transmission of the marked audio data. Moreover, in applications like monitoring the real time detection of the decoder is a mandatory precondition (see section 3.5.2 on page 49).

Adjustability of the algorithm Depending on the application, an adjustment of the operation point between different payload rates and robustness levels should be possible.

Embedding several watermarks Especially in a distribution scenario it is desirable to enable the copyright protection and monitoring at the same time in order to satisfy all the security needs of both the copyright owner and his publisher. As a result the simultaneous embedding of multiple watermarks is required, each of them could fulfill a different requirement (see [80], [21]). Moreover the possibility of embedding more than one watermark at the same time can be used to embed reference signals for improving the robustness of the watermarking algorithm.

Blind Watermarking Furthermore, in application scenarios like the tracing of unauthorized copies or the monitoring of broadcast radio programs it is also necessary to extract information from the watermarked audio data. The required systems are the so-called *blind watermarking* systems presented in section 2.2.2 on page 23. They present the greatest challenge in the development of a watermarking method, since neither the original audio track nor the watermark are used during the detection process.

6.2 Algorithm Design Criteria

The requirements detailed in section 6.1 describe the maximum sets of criteria a watermarking algorithm has to fulfill. The described features like **quality, robustness, security** and **application criteria** can in general not be fulfilled simultaneously in each imaginable application and are strongly related to each other. According to the intended application of the watermarking algorithm different variations and corresponding design criteria are relevant for the development of an effective method. Each application represents one operating point in the 3D surface[1] describing the interdependence of quality, data rate, and robustness of watermarking algorithms (see figure 6.1). The figure outlines the reprocial relationship between

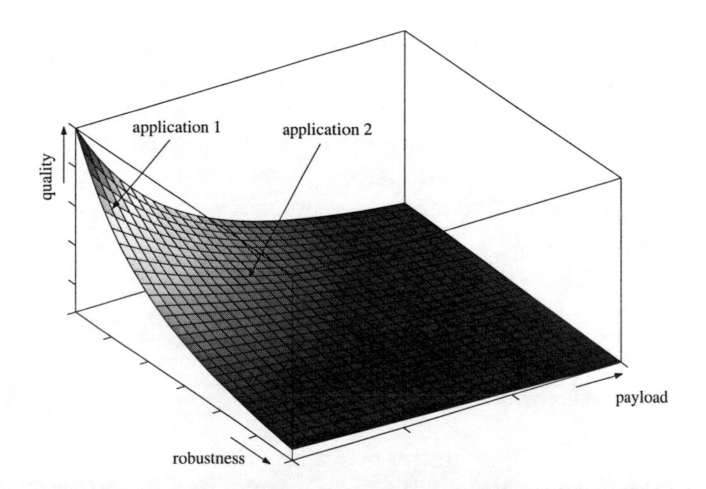

Figure 6.1: Quality, robustness, data rate and their interdependence

the different requirements (see sections 6.3.4 and 6.6.1). The most important requirement addresses the quality of the watermarked

[1]Whereby the surface need not be continuous.

items. If the quality of the audio tracks cannot be preserved, neither users (whether consumers or broadcast industry professionals) nor especially the recording industry will accept this technology. Consequently the definition of robustness of the watermark depends on the quality. A watermark is robust if removal from the audio track results in an implicit and inevitable decrease in quality while at the same time that rendering the audio track unusable for the intended application. This emphasizes the priority in ranking among the requirements from quality as follows:

1. quality
2. robustness and security
3. payload and application criteria

To ensure the quality of the watermarked audio tracks, a psychoacoustic model has to be an integral part of the watermark encoder. This enables the calculation of perception thresholds in order to shape the watermark noise according to psychoacoustic criteria of the human auditory system. After presenting the basic algorithm in the next section and several extensions in section 6.4, the integration of the psychoacoustic model are presented in section 6.5 as one of the essential building blocks of an effective audio watermarking algorithm.

6.3 The Basic Algorithm

In order to develop a watermarking algorithm one must take into account the features of the data as well as the possible modifications due to anticipated manipulations and intended attacks. A typical audio signal in CD quality is represented by a 16-bit sample sequence (16 bps) per channel (two for stereo) with a sample frequency of 44.1 kHz. Accordingly every second carries 88 200 samples or 178 kbyte of data. This large amount of data offers the possibility of using statistical methods relying on large sets. Statistical techniques

implicitly satisfy most of the security requirements of a watermarking algorithm mentioned above. These algorithms are working on distributions, which are generated and defined by means of a secret key, have a mathematical formulation and can be modified and adjusted by parameters to control robustness, quality and payload size. Due to the fact that most of the signal processing like lossy compression, filtering, adding noise, etc. can be described in the frequency domain, it seems to be the most appropriate working domain for watermarking. The marking in the frequency domain also spreads the watermark over the whole data set in the time domain and therefore is less affected by cropping operations. Of course not all of the manipulations imaginable can be described in an easy way in the frequency domain. Time-scaling attacks introduced either intentionally or due to DA/AD-conversions e.g. may result in synchronization problems in the frequency domain.
The basic algorithm described here is a statistical method based on hypothesis testing [2]. It splits the audio track into time slices (frames) for embedding the individual bits of the watermark. The number of samples per frame is a multiple of 512 samples, representing a block. The basic algorithm is a so-called *semi-blind watermarking* method

$$D_K(\hat{\mathbf{c}}_\mathbf{w}, \mathbf{w}) = \hat{\mathbf{w}} \quad \text{and} \quad C_\tau(\hat{\mathbf{w}}, \mathbf{w}) = \begin{cases} 1, & c_\tau \geq \tau \\ 0, & c_\tau < \tau \end{cases} \tag{6.1}$$

Such a method is necessary in applications which cannot assume the original $\mathbf{c_o}$ to be present during the detection process. Furthermore, the semi-blind watermarking method can also be used in copy control applications [89, 21].

6.3.1 Encoding

Method The original signal $\mathbf{c_o}$ is split into $N_B = \left\lfloor \frac{l(\mathbf{c_o})}{l(\mathbf{c_o^B})} \right\rfloor$ blocks $\mathbf{c^B_o}_j$, $0 \leq j \leq N_B - 1$ with $l(\mathbf{c^B_o})$ samples. The set of all N_B blocks represents one frame $\mathbf{c^F_o} = \{\mathbf{c^B_o}_0, \ldots, \mathbf{c^B_o}_{N_B-1}\}$ carrying one bit. The algorithm is structured as follows:

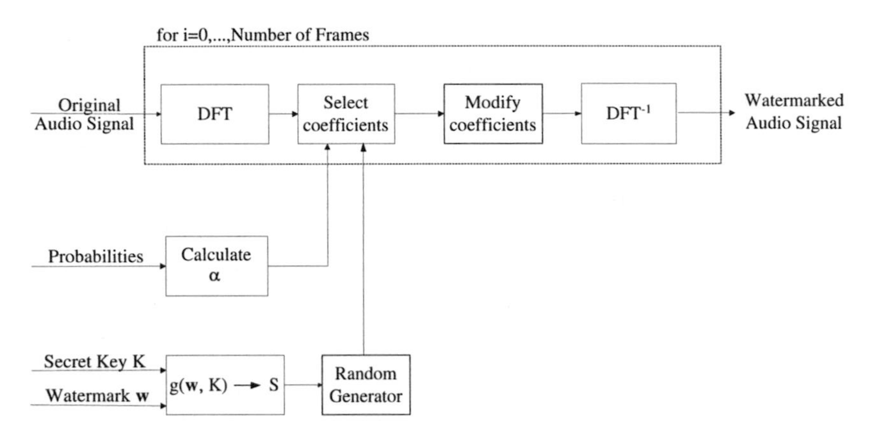

Figure 6.2: Semi-blind watermarking encoder

1. Each block of frame $\mathbf{c_o^F}$ is transformed into the Fourier domain $\mathcal{F}\{\mathbf{c^B_o}_j\}, \forall j$. The magnitudes of the Fourier coefficients are denoted by $\mathbf{C_o^F} = \{\mathbf{C^B_o}_0, \ldots, \mathbf{C^B_o}_{N_B-1}\}$.

2. The secret key K and the watermark $\mathbf{w}$ are mapped to the seed of a random number generator, which generates a pseudorandom sequence **pn**. Based on the pseudo-noise vector **pn** consisting of two equiprobable elements $\{+1, -1\}$, the generator select two subsets C^+ and C^- from set $C = C^+ \cup C^- = \mathbf{C_o^F[i]}, i = 0, \ldots, N-1, N = \frac{N_B}{2} l(\mathbf{c_o^B})$ consisting of the magnitudes of Fourier coefficients (see figure 6.2). The selection results in two subsets, which are of equal size $(|C^+| = |C^-| = M \leq N)$ with no common elements $C^+ \cap C^- = \emptyset$. According to the statistic theory the subsets C^+ and C^- are drawn from the same population and therefore share the same properties, i.e. same mean value and variance of the distribution. This is the basic assumption of this watermarking algorithm, which was first presented in the image watermarking field [14].

$$\mu_o^+ \approx \mu_o^-, \hat{\sigma}_{\mu_o^+} \approx \hat{\sigma}_{\mu_o^-} \tag{6.2}$$

3. A *test hypothesis* H_0 and an *alternative hypothesis* H_1 is formulated. The appropriate test statistic z is defined as a function of the extracted sets $\mathbf{C}^+$ and $\mathbf{C}^-$ which consists of the original $\mathbf{C}_\mathbf{o}^\mathbf{F}$ or marked $\mathbf{C}_\mathbf{w}^\mathbf{F}$ Fourier coefficients. Therefore the random variable z follow two different distributions $f(z|H_0)$in the original and $f(z|H_1)$in the marked case, which is differentiated by comparison against a threshold τ.

$$z = C_\tau(\mathbf{C_o}, \mathbf{pn}) \quad (\textit{original} \text{ audio signal}) \qquad (6.3)$$

$$z = C_\tau(\mathbf{C_w}, \mathbf{pn}) \quad (\textit{marked} \text{ audio signal}) \qquad (6.4)$$

The hypothesis test can now be formulated:

H_0 : If the test statistic follows the distribution $f(z|H_0)$, the audio track carries **no** watermark.

H_1 : If the test statistic does not follow the distribution $f(z|H_1)$, the audio data carries a watermark.

According to figure 6.3 two kinds of errors are incorporated in hypothesis testing:

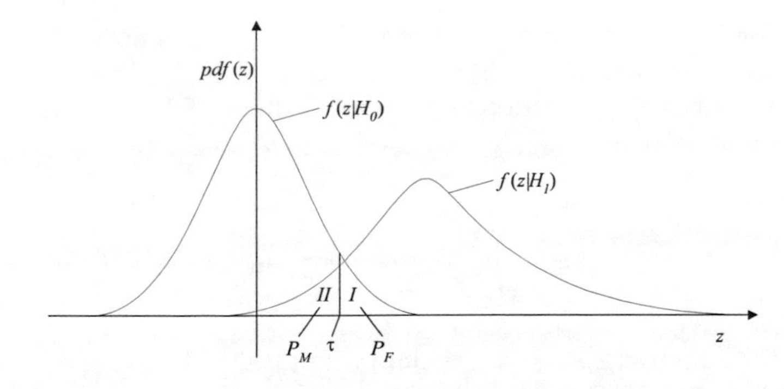

Figure 6.3: Distributions and error probabilities

$$\text{I} : \int_{\tau}^{+\infty} f(z|H_0)dz = P_F \qquad \text{(Type I error)} \tag{6.5}$$

$$\text{II} : \int_{-\infty}^{\tau} f(z|H_1)dz = P_M \qquad \text{(Type II error)} \tag{6.6}$$

P_F is the conditional probability for false alarm and P_M the conditional probability for missing the detection. The detection process is based on the calculation of the test statistic z against the *critical value* or threshold τ. The threshold in turn is derived from the desired errors according to the application. This requires the knowledge of the distribution functions $f(z|H_0)$ and $f(z|H_1)$ in advance.

4. The false detection probability P_F is defined and the threshold τ calculated from equation (6.5).

5. The selected elements $\mathbf{C_o^+}$ and $\mathbf{C_o^-}$ are modified according to the embedding functions e_+, e_-:

$$\mathbf{C_w^+} = e_+(\mathbf{C_o^+}, \alpha), \quad \mathbf{C_w^-} = e_-(\mathbf{C_o^-}, \alpha) \tag{6.7}$$

For a given missing probability P_M and threshold τ, the embedding strength of $\boldsymbol{\alpha}[i] = \alpha, i = 1, \ldots, 2M$ is calculated, which describes the alterations for every element of $\mathbf{C_o^+}$ and $\mathbf{C_o^-}$ from equation (6.8):

$$\int_{-\infty}^{\tau} f(z|H_1)dz = \int_{-\infty}^{\tau} f(C_\tau(\mathbf{C_w^+}, \mathbf{C_w^-}, \mathbf{pn})|H_1)dz = P_M \tag{6.8}$$

The function $e_{+/-}$ is the crucial part of the embedding process. Obviously, one can calculate the parameter vector $\boldsymbol{\alpha}$ to achieve the desired error probabilities. But this takes into account only the statistical features of the data itself. Therefore the changes have to be distributed in such a way which achieves both inaudibility and an overall change resulting in the desired error probability. This aim may not be reached in all cases since the maximum allowed alterations are data dependent. The Fourier coefficients, used as the samples of the statistic, can have a wide range of values. Therefore the embedding functions should introduce relative changes, which are robust against differences in scale

$$e_+ = (1 + \alpha[i])\mathbf{C}_\mathbf{o}^+[i] \quad , \quad e_- = (1 - \alpha[i])\mathbf{C}_\mathbf{o}^-[i] \tag{6.9}$$

$$\mathbf{C}_\mathbf{w}^+[i] = (1 + \alpha[i])\mathbf{C}_\mathbf{o}^+[i] \quad , \quad \mathbf{C}_\mathbf{w}^-[i] = (1 - \alpha[i])\mathbf{C}_\mathbf{o}^-[i] \tag{6.10}$$

for $i = 1, \ldots, M$.

6.3.2 Decoding

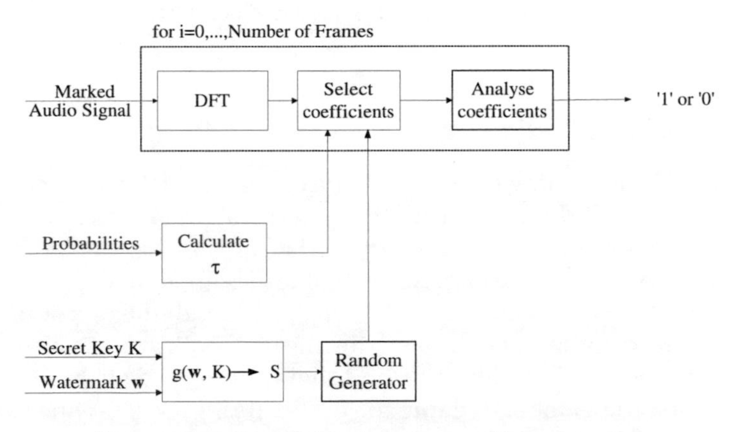

Figure 6.4: Semi-blind watermarking decoder

Method By detecting watermarks the audio signal is transformed in the same way from time domain into the frequency domain. Since the same pseudo random generator is used, the same bit pattern will be produced by the input of same watermarks and keys (see figure 6.4 on page 95). The detection procedure is structured as follows:

1. Map the secret key and watermark to the seed of the random number generator in order to generate the subsets C^+ and C^-. The subsets will contain the altered elements $\mathbf{C}^+_\mathbf{w}[i]$ and $\mathbf{C}^-_\mathbf{w}[i], i = 1, \ldots, M$ in the case of a embedded watermark.

2. The threshold τ for correct detection is defined.

3. To examine the test hypothesis the expectation value $E(z)$ is compared against threshold τ. Thus the test hypothesis can be formulated as follows:

$$H_0 : E\{z\} \leq \tau \quad \text{The hypothesis is accepted} \Longrightarrow \text{watermark } \textbf{not} \text{ embedded}$$

$$H_1 : E\{z\} > \tau \quad \text{The hypothesis is rejected} \Longrightarrow \text{watermark } \textbf{embedded}$$

6.3.3 Test Statistic

A first choice for testing the sample data during the detection process is the mean itself. However during the reading process the detector does not have the knowledge of testing data against a *known* population mean. Therefore the expectation value of both sets C^+ and C^- are compared to examine equation $\mu^+ = \mu^-$. Although the distribution of the underlying set is not known, the mean of both subsets must be same and the distributions of mean follow a normal distribution. A watermark is not embedded into the audio file if $\mu^+ = \mu^-$. Otherwise if $\mu^+ > \mu^-$ the audio data is watermarked.

Test hypothesis and alternative hypothesis

$$H_0 \; : \; \mu^+ = \mu^- \quad \text{watermark } \textit{not} \text{ embedded} \qquad (6.11)$$

$$H_1 \; : \; \mu^+ > \mu^- \quad \text{watermark is } \textit{embedded} \qquad (6.12)$$

Function

$$z_w = C_\tau(\mathbf{C_w}, \mathbf{pn}) = \frac{\mu_w^+ - \mu_w^-}{\sigma_{\mu_w^+ - \mu_w^-}} \tag{6.13}$$

$$= \frac{\frac{1}{M}\sum_{i=1}^{M} \mathbf{C_w^+}[i] - \mathbf{C_w^-}[i]}{\sqrt{\frac{1}{M-1}\left\{\frac{1}{M}\sum_{i=1}^{M}(\mathbf{C_w^+}^2[i] + \mathbf{C_w^-}^2[i]) - (\mu_w^{+2} + \mu_w^{-2})\right\}}} \tag{6.14}$$

Before detecting the watermark from an audio file, it is not known whether it has been manipulated. What is known in advance is the probability distribution function of the marked and unmarked data.

$$H_0 \ : \ f(z|H_0) = N(\mu_0, 1) \tag{6.15}$$

$$H_1 \ : \ f(z|H_1) = N(\mu_1, 1) \tag{6.16}$$

According to the Central Limit Theorem, the distributions of $f(z|H_0)$, $f(z|H_1)$ are both normal with variance $\sigma_z^2 = 1$ and different mean values μ_0 and μ_1 (see figure 6.5). The relative error of the average values

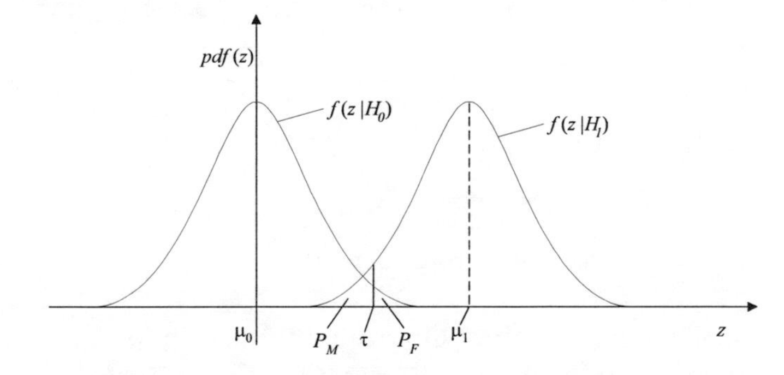

Figure 6.5: Normal distributions for test statistic

ε can be described with the definition of the variation coefficient [64]:

$$\varepsilon := \frac{\hat{\sigma}_{\bar{X}}}{\bar{X}} = \frac{\hat{\sigma}_{\frac{\mu^+ + \mu^-}{2}}}{\frac{\mu^+ + \mu^-}{2}} = \frac{\hat{\sigma}_{\mu^+ + \mu^-}}{\mu^+ + \mu^-} \tag{6.17}$$

With the definition of ε, $q := \frac{\mu^+ - \mu^-}{\mu^+ + \mu^-}$ and equation (6.2) the random variable z can be written as:

$$z = \frac{\mu_w^+ - \mu_w^-}{\sigma_{\mu_w^+ - \mu_w^-}} = \frac{\mu_o^+ - \mu_o^- + \alpha(\mu_o^+ + \mu_o^-)}{\sqrt{(1+\alpha^2)\sigma^2_{\mu_o^+ + \mu_o^-} + 2\alpha(\sigma_o^{+2} - \sigma_o^{-2})}} \quad (6.18)$$

$$= \frac{q_{\mathrm{C_o}} + \alpha}{\sqrt{(1+\alpha^2)\varepsilon^2_{\mathrm{C_o}} + 2\alpha\frac{\sigma_o^{+2} - \sigma_o^{-2}}{(\mu_o^+ + \mu_o^-)^2}}} \quad (6.19)$$

According to the assumptions made in equation (6.2) an approximation for the expected mean value can be derived:

$$\mu_z(\alpha) \approx \frac{\alpha}{\sqrt{1+\alpha^2}} \frac{1}{\varepsilon_{\mathrm{C_o}}}, \quad \mu_0 = \mu_z(0) \approx 0, \quad \mu_1 = \mu_z(\alpha) \quad (6.20)$$

This equation shows the influence of the average value fluctuation of the data on the comparator function defined by equation (6.13). The influence of the data set and the different parameters contained in (6.20) on the error probability will be discussed in the next section.

6.3.4 Calculation of the Error Probability

The error probability P_e is defined by

$$P_e = P_{01}P_0 + P_{10}P_1 \quad (6.21)$$

where $P_{01}P_0$ is the probability for wrongly deciding for H_1 if H_0 is correct, and $P_{10}P_1$ is the probability of wrongly accepting H_0 if H_1 is correct. P_0 and P_1 are the a priori probabilities that the hypotheses H_0 and H_1 are true. $P_{01} = P_F$ is the conditional probability for false alarm and $P_{10} = P_M$ the conditional probability for missing the detection. By using the model of the probability distribution function in the unwatermarked case according to equations (6.15) and (6.16) and assuming a fixed weighting $\boldsymbol{\alpha} := \{\alpha\}_{i=1}^N$ of the pseudo noise sequence the probability distribution function for the two different sequences are

$$f(z|H_0) = \frac{1}{\sqrt{2\pi}} e^{-\frac{(z-\mu_z(0))^2}{2}}, \quad f(z|H_1) = \frac{1}{\sqrt{2\pi}} e^{-\frac{(z-\mu_z(k))^2}{2}} \tag{6.22}$$

Therefore the error probability is obtained by

$$P_e = P_{01}P_0 + P_{10}P_1 = P_0 \int_{\tau}^{+\infty} f(t|H_0)dt + P_1 \int_{-\infty}^{\tau} f(t|H_1)dt \tag{6.23}$$

according to the defined threshold τ. Setting the a priori probabilities that the two hypotheses occur to $P_0 = P_1 = \frac{1}{2}$ and using the definition for the complementary error function $\mathrm{erfc}(x)$

$$\mathrm{erfc}(x) = 1 - \mathrm{erf}(x) = \frac{2}{\sqrt{\pi}} \int_{x}^{+\infty} e^{-t^2} dt \tag{6.24}$$

it can be written with the threshold τ as

$$P_e = P_{01}P_0 + P_{10}P_1 = \frac{1}{2}\left(\frac{1}{2}\mathrm{erfc}\left(\frac{\mu_z(0) + \tau}{\sqrt{2}}\right)\right) + \frac{1}{2}\left(\frac{1}{2}\mathrm{erfc}\left(\frac{\mu_z(\alpha) - \tau}{\sqrt{2}}\right)\right) \tag{6.25}$$

Setting the threshold $\tau = \frac{\mu_z(\alpha)}{2}$ and using the approximations for the different means according to equation (6.20) an approximation for the error probability can be derived:

$$P_e = \frac{1}{2}\mathrm{erfc}\left(\frac{\mu_z(\alpha)}{2\sqrt{2}}\right) \approx \frac{1}{2}\mathrm{erfc}\left(\frac{\alpha}{2\sqrt{1+\alpha^2}}\sqrt{\frac{N}{2}\frac{1}{\epsilon_{C_o}}}\right) \tag{6.26}$$

This equation shows the influence of the original audio track, the design of the underlying algorithm and the effect of embedding the watermark on the error probability. If the randomly selected data set possesses a lower variation coefficient ϵ_{C_o} the shift of the distribution function $f(z|H_1)$ is increased. Therefore, the lower the variation of the original data set around the mean value the better the distinction between the original and the marked audio track for the same amount of modification introduced by the embedding process. The

factor $\frac{\alpha}{\sqrt{1+\alpha^2}}$ relates the embedding strength to the error probability. Increasing the embedding strength by a stronger alteration of the Fourier coefficients decreases the error probability. Furthermore the error probability can be decreased by increasing the number of Fourier coefficients N used for the calculation of the test statistic. This in turn corresponds to the use of a longer time slice for embedding the information.
The watermarking algorithm presented so far embeds one bit into an audio file. This has been extended in order to embed several watermarks consisting of n-bits of information simultaneously into an audio file. The goal has been the extension of the hypothesis test on n-bits.

6.4 Extensions of the Basic Algorithm

6.4.1 Blind Detection

The first question is how to read the embedded watermark from audio data only by using the key. The 1-bit-algorithm differs from n-bit-algorithm primarily regarding the generation of bit pattern. During the embedding process the 1-bit-algorithm codes the watermark $\mathbf{w}$ and the key K by using a hashing function $g(\mathbf{w}, K) = s$. The output of this function s is used as seed for the random number generator and is based on the key and the watermark. In order to be able to detect a watermark one has to decouple the generation of the random number sequence which selects the different subsets C^+ and C^- from the information $\mathbf{m}$ to be embedded. The message $\mathbf{m}$ will be represented by a sequence of $l(\mathbf{m})$ separate symbols, drawn from an alphabet $\mathcal{A}$ of size $|\mathcal{A}|$. The special sequence for the watermark $\mathbf{m}$ to be embedded can be embedded and detected independently. Besides the usage of the '0' and '1' symbol for encoding the bit representation of the message an additional symbol 'S' is used in order to indicate the beginning and the end of the watermark message. Correspondingly three different patterns are generated with the secret key K representing the symbols from alphabet $\mathcal{A} = \{'1', '0', 'S'\}$

of size $|\mathcal{A}| = 3$:

$$p_{('0','1','S')} = \begin{cases} 10101\ldots10, & '0' \text{ pattern} \\ 01010\ldots01, & '1' \text{ pattern} \\ 11000\ldots11, & 'S' \text{ pattern} \end{cases} \quad (6.27)$$

The pattern for each symbol is distributed over one frame of audio data created by concatenating a number of N_B blocks of size $l(\mathbf{c_o}_B) = 512$. The number of samples $l(\mathbf{c_o}_B)$ for the each block is determined by the usage of a psychoacoustic model, which performs a time-frequency analysis on blocks with 512 samples (see section 6.5.1 on page 105). A complete watermark of $l(\mathbf{m})$ bits is embedded in $N_F = l(\mathbf{m}) + 1$ frames, consisting of $l(\mathbf{c_o}_F) = N_B \times l(\mathbf{c_o}_B)$ samples (see figure 6.6). In turn the maximum number of times $N_\mathbf{w}$ the watermark

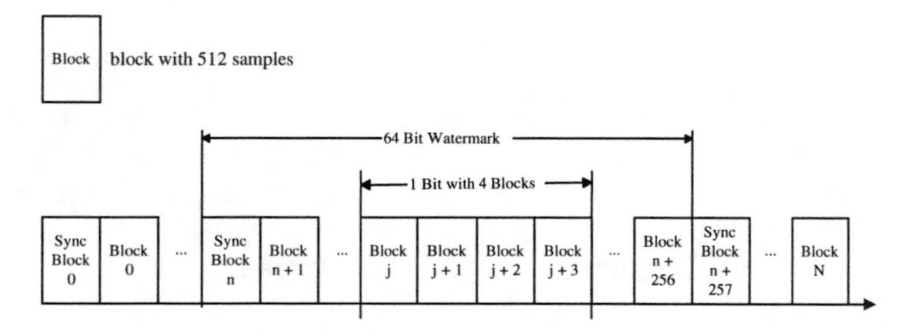

Figure 6.6: Structure of the frame and corresponding blocks

is embedded in the audio stream is

$$N_\mathbf{w} = \left\lfloor \frac{l(\mathbf{c_o})}{(l(\mathbf{m}) + 1) \times l(\mathbf{c_o}_F)} \right\rfloor \quad (6.28)$$

The whole watermark is mapped onto a pattern consisting of the pattern for the individual bits and the synchronization symbol 'S' and embedded in the frames by modifying the selected frequency coefficients (see figure 6.7 on page 102). During the detection process each group of N_B blocks is analyzed by calculating the test function

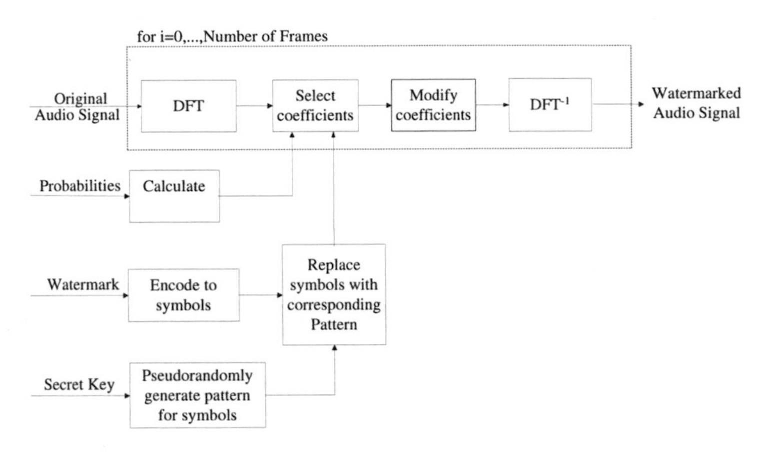

Figure 6.7: Encoding

using the pattern for the different symbols. For a given error probability and the calculated threshold τ the decision about which of the symbols is embedded is performed according to equation (6.13). The complete bit pattern is converted back to the corresponding watermark representation (see figure 6.8 on page 103).

6.4.2 Embedding multiple Watermarks

Watermarks embedded in audio files can serve different purposes. A watermark can be used for the identification of the user or copyright owner,[2] for identification of the track in monitoring applications etc. With the extended algorithm several watermarks are embedded simultaneously into an audio file. The watermarks should not interfere and be spread over the entire audio stream. The modification of the magnitudes in the frequency domain opens up the possibility to divide the spectrum into distinct frequency bands. Each frequency band is intended for the embedding of one watermark [5]. Thereby

[2] Also called fingerprinting.

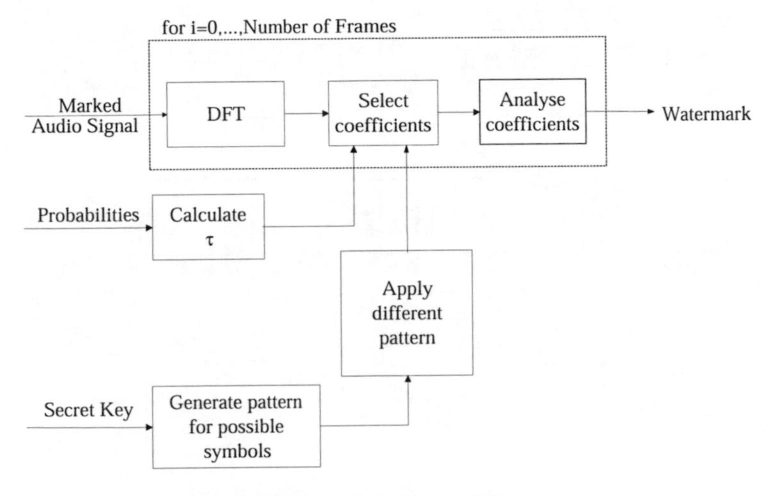

Figure 6.8: Decoding

the orthogonality of the basis function of the Discrete Fourier Transformation (DFT) implicitly avoids that the several watermarks destroy themselves. The following model with three different watermarks divides the spectrum into three frequency bands (see figure 6.9 on page 104):

- A watermark carrying an identification code like the ISRC (International Standard Recording Code) [44]. This is useful in monitoring applications.

- A watermark for user identification.

- A third watermark can be used as a reference signal. This watermark should be first read during the detection process.

The three watermarks can be embedded in one step or sequentially during several processes into the audio track [6].

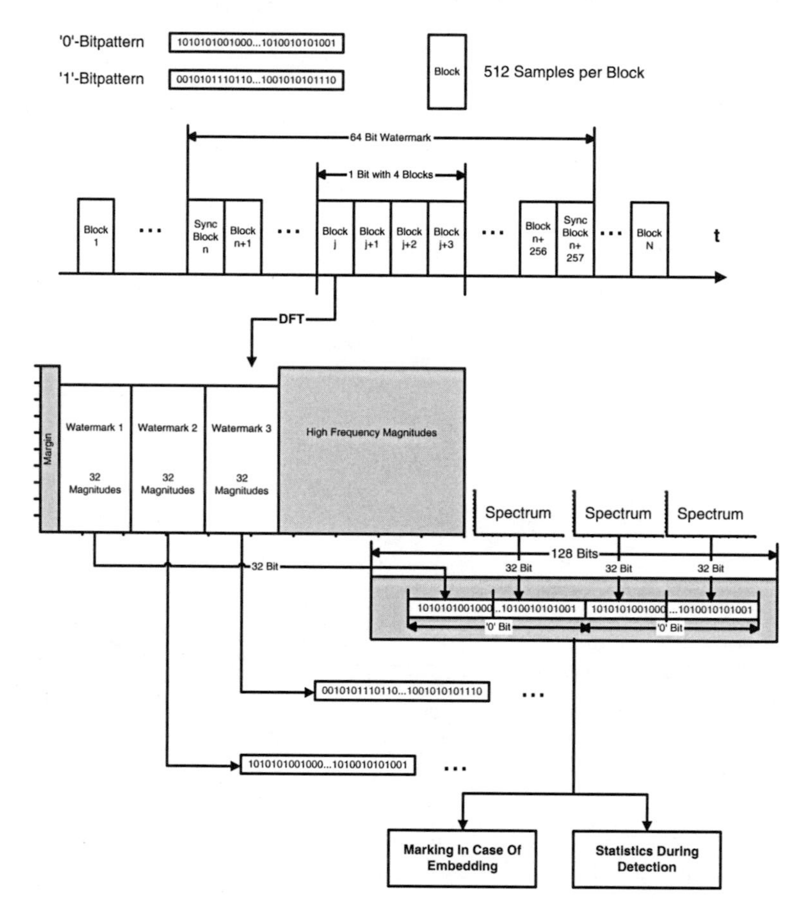

Figure 6.9: Simultaneous embedding of several watermarks

During the embedding process and before the partitioning the spectrum should be examined whether these are suitable for the embedding of a watermark. In the frequency range over 15 kHz the amplitudes have only very small values. Using such amplitudes for a watermark includes the risk that rounding errors cover the relative modification of the amplitude. Furthermore due to the robustness requirement against sampling rate conversion down to $f_s = 22.05$ kHz (see sections 6.1.2 and 8.3.3) frequencies above 11.025 kHz are therefore unsuitable. To increase the robustness the algorithm can be used as semi-blind watermarking system by providing the suspected watermark according to equation (6.1). The pattern of the individual bits are connected to form the bit pattern of the whole watermark. The increased number of coefficients decreases the variance of the probability distribution function of the bit pattern for the whole watermark and therefore decreases the error probabilities.

6.5 Psychoacoustic Watermarking

6.5.1 Psychoacoustic model MPEG 1 Layer I and II

Psychoacoustic models used in current audio compression encoders apply the frequency and temporal masking effects in order to ensure inaudibility by shaping the quantization noise according to the masking threshold. In turn a natural approach is to use already existing models for shaping the watermark noise. The different psychoacoustic models differ in complexity and the implementation of the different masking effects. One of the frequently used models is the psychoacoustic model 1 layer I and II of ISO-MPEG with $f_s = 44.1$kHz [49]. It supports the sampling rates $f_s = 32, 44.1$ and 48 kHz. In order to iteratively allocate the necessary bits the MPEG standard calculates the signal-to-mask ratios (SMR) of all the subbands. This requires the determination of the maximum signal level and the minimum masking threshold in each subband.

Calculation of the power density spectrum

To derive the masking threshold the power density spectrum of the input block has to be estimated. In order to minimize the leakage effect the input block is multiplied with a Hanning window defined by:

$$h(i) = \sqrt{\frac{8}{3}}\left(1 - \cos\left(\frac{2\pi i}{N}\right)\right), i = 0, 1, \ldots, N-1 \qquad (6.29)$$

Layer I uses an input block $s(l), l = 1 \ldots N$ of length $N = 512$ whereas Layer II is operating on blocks with $N = 1024$ samples. After multiplication with the Hanning window the FFT of the input block is performed.

$$X(k) = 10\log_{10}\left|\frac{1}{N}\sum_{l=0}^{N-1} h(l)s(l)e^{-i\frac{2\pi kl}{N}}\right|^2 \text{ [dB]} \qquad (6.30)$$

The maximum value of the power density spectrum X is normalized to a value of +96 dB.

Determination of sound pressure level

The sound pressure level L_s in band n is calculated by

$$L_s(n) = max(X(k), 20 * \log_{10}(\text{scf}(n) * 2^{15}) - 10)\text{[dB]} \qquad (6.31)$$

with $X(k)$ in subband n. $X(k)$ is the result form the power density spectrum calculation. scf(n) is the scaling factor in subband n. After the segmentation of the frequency bands this factor is determined from the maximum value of 12 successive samples in band n via a lookup table. According to that lookup table the scaling factor only determines the peak level for a period of time. The multiplication by a factor of 2^{15} is the normalization to +96dB. The −10 dB term corrects the difference between peak and RMS level (see [49]).

Threshold in quiet

The threshold in quiet LT_q (also called absolute threshold) is defined as the sound pressure level of a pure tone that is just audible as a function of the frequency [109] (see figure 6.10 on page 107). The following expression can be used as an approximation:

$$LT_q = 3.64\left(\frac{f}{\mathrm{kHz}}\right)^{-0.8} - 6.5e^{-0.6\left(\frac{f}{\mathrm{kHz}}-3.3\right)^2} + 10^{-3}\left(\frac{f}{\mathrm{kHz}}\right)^4 \text{ [dB]} \quad (6.32)$$

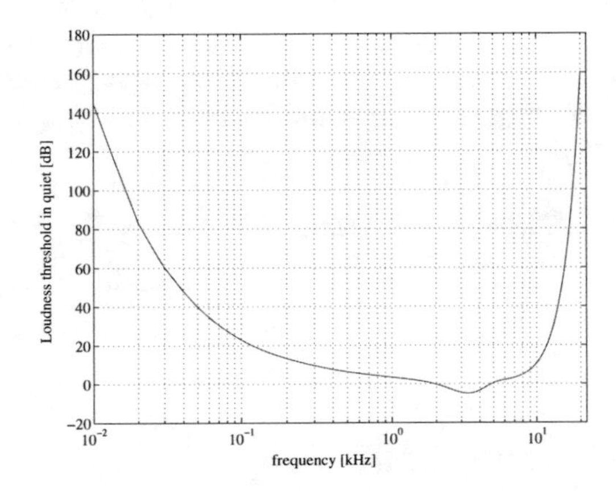

Figure 6.10: Threshold in quiet.

The normalization of the absolute threshold in quiet is done by adjusting the function according to the following rule: a signal with a frequency of 4 kHz and an amplitude of ±1 LSB [3] lies on the curve of the absolute threshold [108].
The absolute threshold is available in form of tables for the different sampling rates f_s (see [49]). In order to take the threshold in quiet into account in the calculation of the global masking threshold the tables contain values for all frequencies necessary to compute the masking threshold. An additional corrective offset has to be added depending on the overall bit rate used per channel.

$$\text{offset} = \begin{cases} -12\,\text{dB, bit rate} < 96\,\text{kbit/s} \\ 0\,\text{dB, bit rate} \geq 96\,\text{kbit/s} \end{cases} \quad (6.33)$$

[3] Amplitude resolution is 16 bit.

If the computed masking threshold lies below the threshold in quiet the masking threshold is set to the absolute threshold in each band.

Determination of tonal and non-tonal components

The masking curves are determined by the tonality of the individual masker (see section 4.2.1 on page 56). Therefore the discrimination between the different components has to be done. The first step is a determination of the local maxima in the power density spectrum (see figure 6.11).

$$X(k-1) < X(k) \text{ and } X(k) \geq X(k+1) \tag{6.34}$$

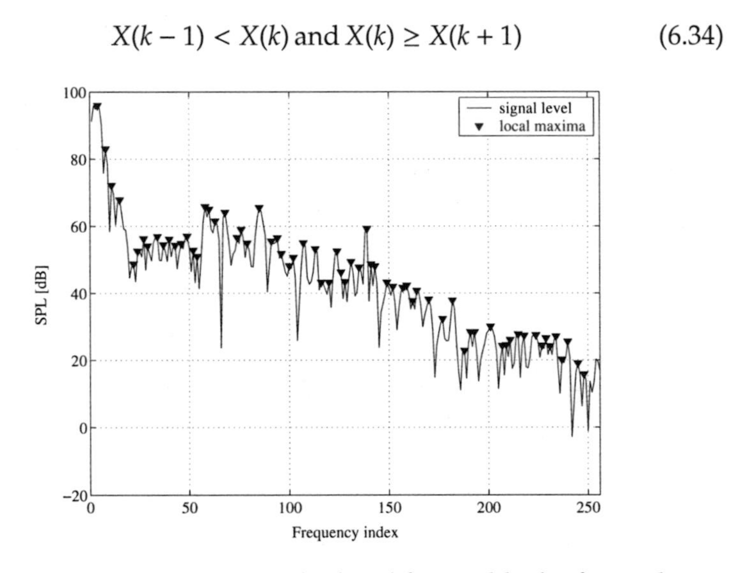

Figure 6.11: Local maxima calculated from a block of samples

Following this step the determination of the tonal components within the bandwidth of a critical band is based on the examination of the power of neighboring spectral lines. A local maxima $X(k)$[4] is also a tonal component if the following criterion is met:

[4] k is the frequency index.

$$X(k) - X(k-j) \geqq 7\,[\text{dB}] \tag{6.35}$$

with

$$\begin{array}{rrlrcccr} j = & -2, +2 & \text{for} & 2 & < & k & < & 63 \\ j = & -3, -2, +2, +3 & \text{for} & 63 & \leqq & k & < & 127 \\ j = & -6, \ldots, -2, +2, \ldots, +6 & \text{for} & 127 & \leqq & k & < & 250 \end{array} \tag{6.36}$$

The sound pressure level of the tonal maskers are computed via

$$X_{tm}(k) = 10 \log_{10}\left(10^{\frac{X(k-1)}{10}} + 10^{\frac{X(k)}{10}} + 10^{\frac{X(k+1)}{10}}\right) [\text{dB}] \tag{6.37}$$

The non-tonal components are computed from the remaining lines without the tonal components within each critical band. The power of these spectral lines is summed to form the non-tonal component $X_{nm}(k)$ [5] corresponding to the critical band. The index k of the non-

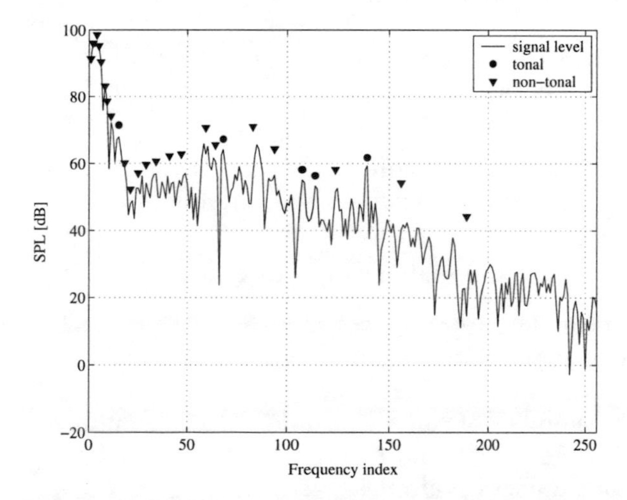

Figure 6.12: Tonal and non-tonal components

[5]The indexes $_{nm}$ and $_{tm}$ denotes the non-tonal respectively the tonal maskers.

tonal component is defined by the index of the spectral line nearest to geometric mean of the critical band. Figure 6.12 on page 109 displays a power density spectrum with the found tonal and non-tonal components for a block of $N = 512$ samples and sampling rate $f_s = 44.1\text{kHz}$.

Decimation of tonal and non-tonal components

The number of maskers considered for the calculation of the global masking threshold is reduced in this step.

- Components are removed from the list of relevant components if their power values are below the absolute threshold.

$$X_{tm}(k) \geq LT_q(k) \text{ or } X_{nm}(k) \geq LT_q(k) \tag{6.38}$$

- For the tonal components an additional decimation is performed if two or more components are less than 0.5 Bark separated. The tonal component with the highest power is kept whereas all other components are removed from the list of tonal components. This operation is performed by applying a sliding window of width 0.5 Bark in the critical band.

The remaining tonal and non-tonal components are used in the calculation of the individual masking thresholds.

Calculation of individual masking thresholds LT_{tm} and LT_{nm}

The MPEG model uses only a subset of the N/2 spectral lines to calculate the global masking threshold. The reduction to the sub-sampled frequency domain is a non-linear mapping of the $N/2$ frequency lines. The number of samples used in the sub-sampled frequency domain is different depending on the sampling rate and layers. For Layer I the number of samples are

$$\begin{array}{ll} f_s = 32\,\text{kHz} & n = 108 \\ f_s = 44.1\,\text{kHz} & n = 106 \\ f_s = 48\,\text{kHz} & n = 102 \end{array} \tag{6.39}$$

The masking thresholds for tonal and non-tonal masker can be calculated with the equations:

$$LT_{tm}[z(j), z(i)] = X_{tm}[z(j)] + \mathrm{av}_{tm}[z(j)] + \mathrm{vf}[z(j), z(i)]\ [\mathrm{dB}] \quad (6.40)$$
$$LT_{nm}[z(j), z(i)] = X_{nm}[z(j)] + \mathrm{av}_{nm}[z(j)] + \mathrm{vf}[z(j), z(i)]\ [\mathrm{dB}] \quad (6.41)$$

The masking threshold is calculated at the frequency index i. j is the frequency index of the masker. $X_{tm}(z(j))$ is the power density of the masker with index j. The term $\mathrm{av}_{t|nm}[z(j)]$ is the so-called masking and $\mathrm{vf}[z(j), z(i)]$ the masking function.

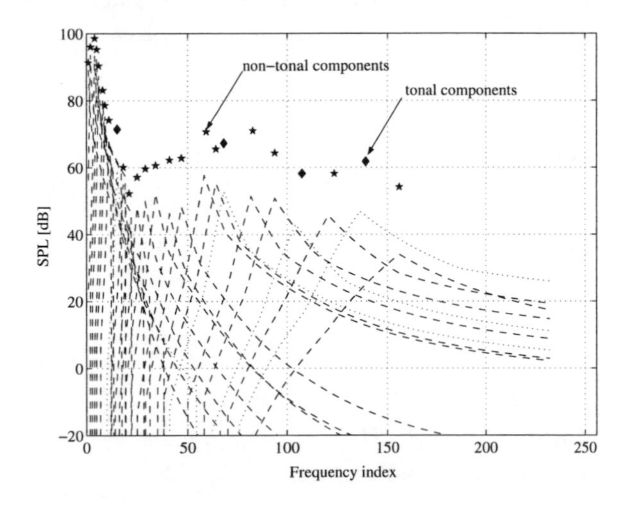

Figure 6.13: Individual masking thresholds for the tonal components

Figure 6.13 displays the decimated tonal masker and the corresponding individual masking thresholds vf. $z(j)$ is the so-called *critical band rate* and denotes the frequency in the Bark scale. The bark values and the corresponding frequency indexes are stored in tables.
The masking index for tonal masker is defined as:

$$\mathrm{av}_{tm}[z(j)] = -1.525 - 0.275 * z(j) - 4.5\ [\mathrm{dB}] \quad (6.42)$$

The non-tonal masking index can be calculated by the formula:

$$\mathrm{av}_{nm}[z(j)] = -1.525 - 0.175 * z(j) - 0.5\,[\mathrm{dB}] \tag{6.43}$$

The masking function $\mathrm{vf}[z(j), z(i)]$ with the distance in Bark $\Delta z = z(i) - z(j)$ is defined by

$$\mathrm{vf} = \begin{cases} 17(\Delta z + 1) - (0.4X[z(j)] + 6) & -3 \leqq \Delta z < -1 \\ (0.4X[z(j)] + 6) * \Delta z & -1 \leqq \Delta z < 0 \\ -17\Delta z & 0 \leqq \Delta z < 1 \\ -(\Delta z - 1) * (17 - 0.14X[z(j)]) - 17 & 1 \leqq \Delta z < 8 \\ \text{in [dB]} & \text{in Bark} \end{cases} \tag{6.44}$$

Calculation of the global masking threshold LT_g

In order to calculate the global masking threshold LT_g the different components have to be summed up. The global masking thresholds for frequency index i are computed by adding the power of the threshold in quiet and the tonal and non-tonal masker in each case.

$$LT_g(i) = 10\log_{10}\left(10^{\frac{LT_q(i)}{10}} + \sum_{j=1}^{m} 10^{\frac{LT_{tm}[z(i),z(j)]}{10}} + \sum_{j=1}^{n} 10^{\frac{LT_{nm}[z(i),z(j)]}{10}}\right) [\mathrm{dB}] \tag{6.45}$$

Figure 6.14 on page 113 displays the global masking threshold as calculated according to equation 6.45.

Computation of the minimum masking threshold LT_{Min}

The global masking threshold LT_g is computed in the sub-sampled frequency domain with the number of spectral lines according to table 6.39 on page 110. This frequency indexes are mapped on to the 32 subbands.

$$LT_{Min}(n) = \min_{f(i) \text{ in subband } n} [LT_g(i)]\,[\mathrm{dB}] \tag{6.46}$$

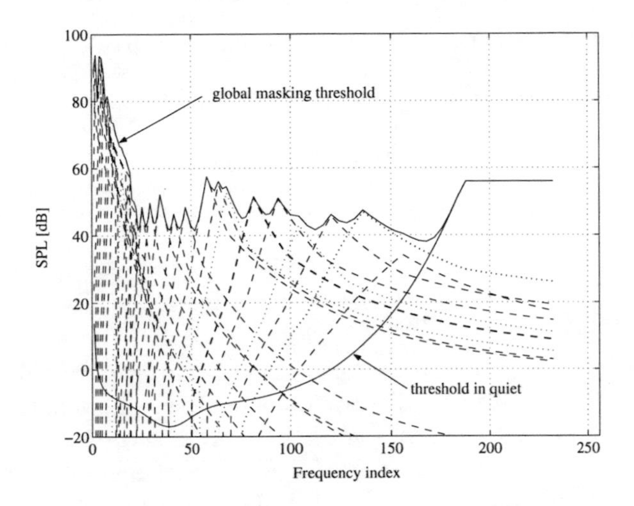

Figure 6.14: Global masking threshold

Calculation of the signal-to-mask ratio

The SMR is calculated for every subband n

$$SMR(n) = L(n) - LT_{Min}(n)\,[\text{dB}] \tag{6.47}$$

6.5.2 Integration into Watermark Embedder

The output of the psychoacoustic model is the signal-to-mask ratios (SMR). This information is used by lossy audio compressors to iteratively allocate the bits in every subband. This is not necessary in the case of a watermarking application, since only the masking threshold for each block is of interest. Consequently the integration of the psychoacoustic model requires only the following steps:

1. Calculation of the power spectrum.
2. Identification of the tonal (sinusoid-like) and non-tonal (noise-like) components.

3. Decimation of the maskers to eliminate all irrelevant maskers.

4. Computation of the individual masking thresholds.

5. Computation of the global masking threshold.

6. Determination of the minimum masking threshold in each sub-band.

Perceptual watermark encoder

The task of a watermark encoder is to shape the watermark in order to ensure inaudibility on the one hand and simultaneously embed the watermark with the maximum power according to the carrier signal to provide maximum robustness. The encoder consists of several components (see figure 6.15). The noise generator produces

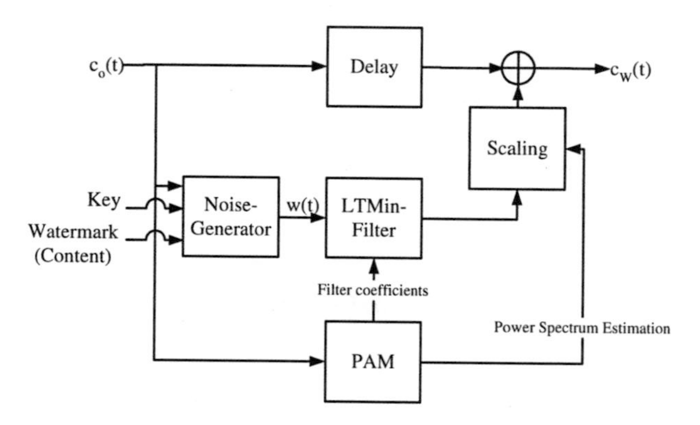

Figure 6.15: Watermark encoder and its components

noise by means of a secret key *Key*. The psychoacoustic model (PAM) block analyzes the original signal $s(t)$ in order to calculate the minimum masking threshold LT_{Min}. This threshold defines the frequency response of the spectral weighting process, which shapes the watermark. The shaped watermark signal is sometimes scaled in order to shift the watermark noise below the masking threshold. The result is the actual watermark signal added to the original signal to produce the watermarked track $s_W(t)$.

Watermark noise generator The task of a noise generator is to produce a signal based on the watermark information and a key, that must not be distinguishable from random noise. This watermark is the input of the spectral shaping block. In a more general solution information from the original signal like the phase and/or amplitude can be taken into account. The signal can undergo a transformation T in another domain (Fourier, Wavelet, Cepstrum, etc.) where the embedding is performed.

Spectral weighting of watermark signal The spectral weighting block multiplies the frequency components of the watermark noise with the weight coefficients calculated by the PAM block. The LT_{Min} filter block is based on a piecewise linear approximation of the masking threshold which represents the frequency response. If a m-tap finite impulse response (FIR) filter (with m even) is used the original has to be delayed by $m/2$ samples (see Delay block in figure 6.15) before added to the watermark noise. This adaptive filtering of the watermark noise ensures the inaudibility of the watermark noise based on the psychoacoustic features. The filter used in approximating LT_{Min} can be a FIR filter designed with the window method. A well known window is the Hanning one:

$$w(k+1) = \alpha - (1-\alpha)\cos\left(\frac{2\pi k}{N-1}\right), \alpha = 0.54, k = 0, 1, \ldots, N-1 \quad (6.48)$$

The signal loss due to the window edges can be minimized by overlapping successive blocks by 50% (see [18]). These calculations have to be performed for each block of length $\Delta t = \frac{512 \text{ samples}}{f_s \text{ samples/s}}$ with sampling frequency f_s. The masking threshold according to MPEG 1 Layer I is calculated for 32 subbands. The reduction to 32 subbands is not necessary in the spectral weighting process. The frequency resolution can be made finer in order to increase the power and robustness of the embedded watermark.

Adjustment of watermark noise level After filtering the watermark noise the sound pressure level has to be adjusted. To calculate the correct attenuation of the watermark noise the power spectrum is estimated. The masking threshold is the result from the psychoacoustic analysis. It is based on a normalized original signal with a maximum shifted to +96dB. This normalization has to be taken into account in the scaling block if one calculates the power of the watermark noise relative to the original signal (see figure 6.16). The

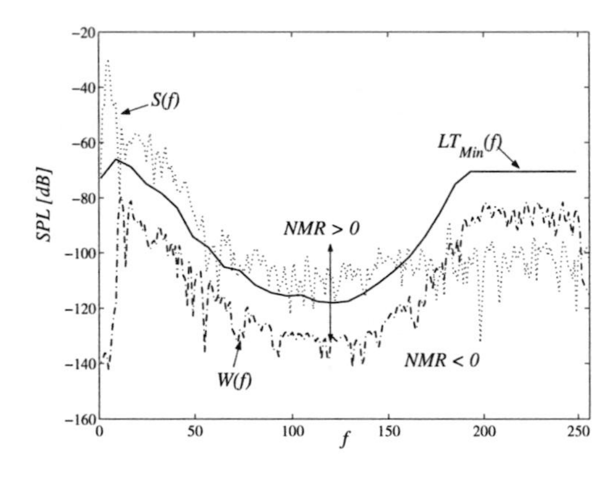

Figure 6.16: Sound pressure levels

calculated attenuation factor is used to adjust the watermark noise power according to equation

$$W_{Noise} = LT_{Min} + NMR\,[\text{dB}] \tag{6.49}$$

The so-called noise masking ratio NMR serves as additional attenuation factor which adjusts the level of quality against robustness.

6.6 Robust Audio Watermarking

The robustness of the watermark can be improved in several ways. At the encoder site the strength of embedding can be increased by altering the psychoacoustic model which will be presented in the next section. Improvements at the decoder site can be made by using the test statistic as a robustness measure as presented in section 6.6.2.

6.6.1 Improvement at the Encoder

According to the presentation of the psychoacoustic model in section 6.5.1 the calculation of the masking threshold $LT_{Min}(n)$ is done for each subband n of the frequency spectrum:

$$LT_{Min}(n) = \min_{f(i)\in n} LT_g(i)\,[dB], \quad n = 1,\ldots,32, \quad i = 1,\ldots 106 \qquad (6.50)$$

It is the minimum of the local masking threshold and the threshold in quiet $LT_q(i)$[6] in the corresponding subband n. The 32 subbands are the result of a reduction of the 106 frequency indexes (for Layer I and sampling rate $f_s = 44.1kHz$) with a nonlinear frequency resolution. In order to maximize the power of the embedded watermark the minimum relation in equation (6.50) should be lowered. To investigate the effect the implementation of the PAM was changed to allow different number of subbands in the frequency domain. For a comparison to the relative alterations defined by a fixed embedding factor α according to equations (6.18) and (6.20), an effective embedding strength is defined via the equation:

$$\Delta\mu_o^+ + \Delta\mu_o^- \;:=\; \alpha(\mu_o^+ + \mu_o^-)\,,\Delta\mu_o^{(+|-)} = \frac{1}{M}\sum_{i=1}^{M} \boldsymbol{\alpha}[i]\mathbf{C}_o^{(+|-)}[i] \qquad (6.51)$$

$$\alpha \;=\; \frac{\Delta\mu_o^+ + \Delta\mu_o^-}{\mu_o^+ + \mu_o^-} \qquad (6.52)$$

[6] i is the index of the sub-sampled frequency domain and tabulated in the ISO standard.

The effective embedding strength can be determined for varying number of subbands and different audio files (see figure 6.17). α describes the relative variation of all Fourier coefficients used for embedding.

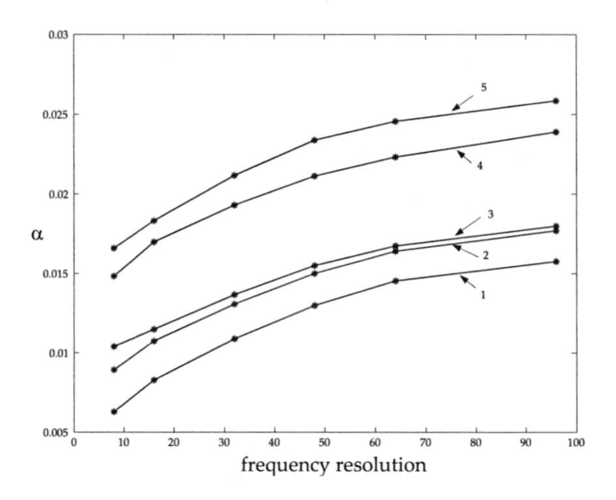

Figure 6.17: Effective factor as function of filter resolution for different tracks ($1-5$).

According to the results presented above a higher frequency resolution corresponds to a higher effective embedding strength. Despite the fact that individual audio tracks have different embedding factors the run of the curve is the same.

6.6.2 Robust Detection

One of the most important preconditions for the A©WA algorithm is a synchronization of the detection window with the embedding one. Attacks like cropping or a time shift of the whole audio file misalign the detector with the embedded watermark respectively the pseudo-random sequence with the corresponding Fourier coefficients. This can happen during attacks like the randomly cropping and inserting of samples or even a time slice of the audio track. Moreover filtering process produce a delay of the samples which results in a time shift of the whole audio track. This can happen intentionally by applying a filtering attack (see section 6.1.2 on page 85) or unintentionally if the audio track is compressed before delivery. All lossy compression algorithms [76] perform some kind of time-frequency analysis of the samples in the time domain. They use filter banks, which divide the signal spectrum into frequency subbands used for perceptual noise shaping (see section 6.6.1 on page 117). As a result of this de-synchronization the values obtained from the comparator function (see equation 6.13 on page 97) decrease to a value which is comparable to the unmarked case. In turn the algorithm described in this section synchronizes the embedding with the detection window using the test statistic as a quantative measure.

Synchronization Method The aim of the synchronization is to align the embedding and detection frame. In this case the pattern is mapped to the correct Fourier coefficients modified during embedding in order to compute the value of the test statistic (6.13) for the different embedded symbols. Since the pattern of one symbol is distributed over one frame consisting of N_B blocks with length $l(\mathbf{c}_{\mathbf{w}B})$ (see figure 6.9 on page 104), the number of blocks $M = N_B \times (N_F + 1)$ is determined by the parameter N_F, the number of frames used during synchronization. The additional frame is used for the shifting procedure to perform the frame adjustment as described in the following. The detection procedure has to be performed in the Fourier domain, which requires the DFT of all M blocks prior to the calculation of the test statistic (6.13) for the N_F frames. After the maximum number of $l(\mathbf{c}_{\mathbf{w}B}) - 1$ search steps in the time domain a detection block is shifted over the whole embedding block within a frame. Correspondingly for one specific search step the blocks are correctly aligned which is in general not the case for the frames. In each search step synchronization of the frames is tested by shifting the different part of the pattern distributed over the N_B blocks $N_B - 1$ times. The calculation

Algorithm 1 Synchronization algorithm.

1: {Input are the $l(\mathbf{c_{w}}_B) \times N_B \times (N_F + 1)$ samples $\mathbf{c_{w}}_S$ of the audio stream}
2: $N = N_B \times N_F$
3: {Search through one block of samples.}
4: **for all** $i < l(\mathbf{c_{w}}_B) - 1$ **do**
5: {Calculate DFT of blocks starting at sample position i.}
6: **for all** $k < N$ **do**
7: $\mathbf{C}^{\mathbf{B}}_{\mathbf{w}k} = \mathcal{F}(\mathbf{c}^{\mathbf{B}}_{\mathbf{w}k})$
8: **end for**
9: {Shift start position by one block.}
10: **for all** $j < N_B - 1$ **do**
11: $\mathbf{C}^{\mathbf{S}}_{\mathbf{w}} = \{\mathbf{C}^{\mathbf{B}}_{\mathbf{w}j}, \ldots, \mathbf{C}^{\mathbf{B}}_{\mathbf{w}N}\}$
12: {Calculate maximum statistic value for all symbols $\in \mathcal{A}$ and frames $\mathbf{C}^{\mathbf{F}}_{\mathbf{w}}$.}
13: **for all** $m < N_F$ **do**
14: $z(i, j, m) = \max_{k=1,\ldots,|\mathcal{A}|} C_\tau(\mathbf{C}^{\mathbf{F}}_{\mathbf{w}m}, \mathbf{pn}_k)$
15: **end for**
16: **end for**
17: **end for**
18: {Calculate the average comparator value of all N_F frames.}
19: **for all** $i < l(\mathbf{c_{w}}_B) - 1$ **do**
20: **for all** $j < N_B - 1$ **do**
21: $z(i, j) = \frac{1}{N_F} \sum_{m=0}^{N_F-1} z(i, j, m)$
22: **end for**
23: **end for**
24: {Find search step index i and block index j with yield maximum value in matrix.}
25: $\{(i, j) | z(i, j) = \max z(i, j)\}$
26: {Calculate synchronization position.}
27: $syncPos = i + j * l(\mathbf{c}^{\mathbf{B}}_{\mathbf{w}})$
28: return $syncPos$

of the comparator function is performed for $(l(\mathbf{c_{w}}_B) - 1) \times (N_B - 1)$ combinations for N_F frames and $|\mathcal{A}|$ possible pattern embedded. The resulting values are stored in a $(l(\mathbf{c_{w}}_B)-1)\times(N_B-1)\times|\mathcal{A}|\times N_F$ matrix. The speed of the whole synchronization procedure can be increased by decreasing N_F and $l(\mathbf{c_{w}}_B) - 1$ respectively the step size Δ for fixed N_B and $|\mathcal{A}|$. Having calculated the matrix, the maximum values of the different patterns are searched in the third dimension of the $(l(\mathbf{c_{w}}_B) - 1) \times (N_B - 1) \times |\mathcal{A}|$ matrix for the frames $i = 1, \ldots, N_F$. The resulting $(l(\mathbf{c_{w}}_B) - 1) \times (N_B - 1)$ submatrices are averaged over the N_F frames. This averaging makes the procedure robust against frames containing bad statistic values. In turn using a larger N_F makes the synchronization more robust at the cost of a higher number of calculations. The indices of the maximum value in the resulting matrix are used for the determination of the synchronization procedure. The general synchronization algorithm for a given $N_F, N_B, |A|$ and a step size $\Delta = 1$ is illustrated by Alg. 1 on page 120. The result of the

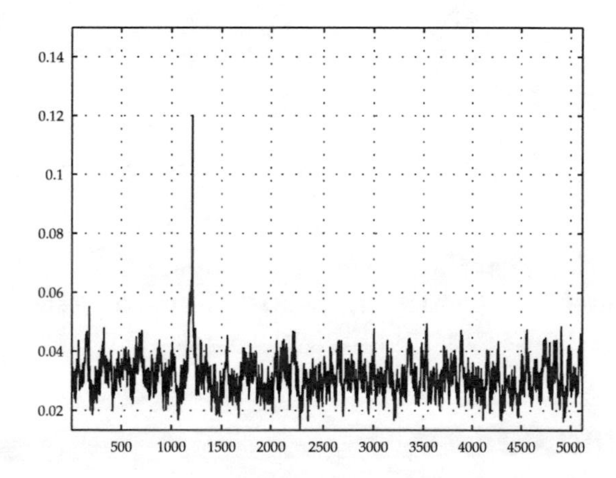

Figure 6.18: Averaged values over one frame of samples

synchronization procedure can be seen in figure 6.18, which shows the averaged statistic values as a function of the samples in the time domain for an audio track compressed at a bit-rate of 128 kBit/s. The peak at sample position 1201 indicates the exact synchronization position. Using a larger step size $\Delta > 1$ increases the speed of the synchronization process, but also the probability of missing the exact synchronization position. This is verified by the sharp peak around the exact synchronization position in figure 6.19. A good

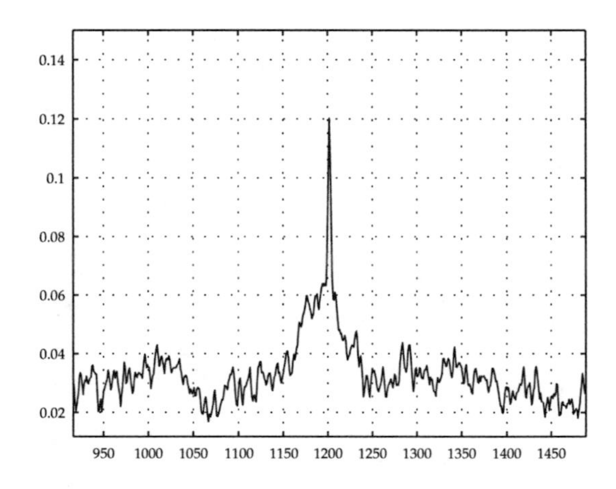

Figure 6.19: Averaged values around the synchronization position

compromise between reliable synchronization results and the speed of the synchronization procedure has been found for a step size $\Delta = 3$ and $N_F = 10$ frames corresponding to one second of audio material.

6.7 Summary

This chapter developed the algorithm for watermarking digital audio data being the main focus of this dissertation. The design process was based on a detailed categorization of the different requirements into quality, robustness, security and application specific features and their interdependence. From these considerations a ranking among the requirements from quality (first) to robustness (second) and data capacity (third) influencing the design process was discussed. Therefore the development of this method was driven on the one hand on the optimization of the parameters and on the other hand on the possibility to easily adjust the algorithm for the variety of application scenarios. The basic algorithm described relys on a statistical model based on decision theory to decide about the presence of a watermark and was motivated by the large amount of data representing digital audio data. Since applications like monitoring require watermarking systems to be able to detect the watermark without the original an corresponding extension of the algorithm was presented. Furthermore the simultaneous embedding of several non-interfering watermarks serving different purposes like copyright owner and customer identification was demonstrated. Preserving the quality of the watermarked audio tracks and the robustness of the embedded watermarks were addressed in two subsequent sections. The quality of the marked tracks is ensured by the presented integration of a psychoacoustic model which controls an adaptive filter shaping the watermark noise according to the masking behavior of the human auditory system. Several improvements of the robustness of the embedded watermarks are described both on encoder and decoder site to increase the power of the embedded noise and the performance of watermark detection. In order to measure the achieved quality of the watermarked tracks and robustness of the watermarks against manipulations and attacks a detailed evaluation is reported in the next two chapters.

CHAPTER
SEVEN

Quality Evaluation of Watermarked Audio Tracks

As is the case for the security issues of the watermarks and corresponding algorithms, the quality of the watermarked work and the robustness of the watermarks have to be defined and approved for each application. Quality evaluation is of considerable importance since the usefulness of digital content is mainly defined by the quality which is presumed for the intended application. Methods for evaluating the quality of watermarked objects are detailed in this chapter. The problem associated with the quality evaluation of watermarked audio data is presented in the first section 7.1. This is followed by a section 7.2 discussing subjective evaluation methods ranging from classical psychophysical methods to current standards and their application to evaluating the distortion introduced by the watermarking process.

Since subjective listening tests are time consuming, expensive and dependent on many not easily controllable parameters, objective methods for determining the quality of watermarked tracks are discussed in section 7.3. Nevertheless certain applications may distribute watermark items in reduced quality. Therefore the following section 7.4 discusses methods currently under standardization which are applicable for evaluation of items with reduced quality and summarizes the different evaluation methods as a function of the intended quality of watermarked audio tracks. The last section 7.5 of this chapter presents the results of subjective listening tests, objective measurement and the comparison of the different methods against each other and the quality of current high quality audio compression coders.

7.1 Quality Evaluation of Watermarked Objects

Unfortunately the output quality of a watermarking codec cannot be quantified easily in objective terms. This is especially a problem for evaluating the watermarking methods operating on new data types like 3D models or images of music scores, since quality criteria and corresponding evaluation procedures are currently not known and therefore subject of current research efforts.
On the other hand, this problem is very similar to the task of evaluating the quality of perceptual coders in the audio field. The research in these fields is driven by the development of highly effective data compression software preserving the quality at the same time. Distortions introduced during perceptual coding are due to quantization noise added in the coding process.
The quantization noise is hidden below the computed perceptual threshold. In watermarking systems, an additional signal carrying information is added likewise. This signal is shaped according to the masking threshold to ensure the quality of the watermarked signal. Conversely, the problems related to the quality of the watermarked carrier signal are expected to be very similar to the perceptual coding case. Therefore an evident approach is to use principles and test procedures already investigated and applied during the development of the compression algorithms. In general two kind of tests are applied. Human measurement techniques and objective evaluation methods.

7.2 Subjective Quality Evaluation

In this context subjective quality evaluation tests are still used as a tool for codec quality evaluation. Standardized test procedures have been developed to maximize the reliability of the results of subjective testing. The next section describes general methods developed to perform subjective quality testing which can be applied regardless of the media type. Subsequently subjective and if available corresponding objective quality measurement techniques are detailed for perceptual audio, image and video coders. This includes approaches to adapt these techniques to the watermarking problem.

7.2.1 Psychophysical Methods

Subjective evaluation can be performed in two different ways. Testing the transparency of the watermarked items or in a more general way rating the quality of the processed items with respect to the reference signal. In the following, an item is called *transparent* if no differences between the original and the watermarked version are perceivable. Otherwise it is called *non-transparent*.
All tests involving human beings have in common that they require

- a specification of the evaluation environment
- a careful selection of the test material
- a training phase for the assessors of the test
- a test phase which consists of a comparison of the coded material against a reference (original)
- a statistical analysis for a quantative interpretation of the results.

The field of psychophysics[1] is the science to derive correlation between quantative variables and qualitative experience of human beings. Corresponding psychophysical methods are used wherever

[1] A term coined by the German physicist and psychologist Gustav Fechner (1801-1887).

studies involving the judgment of subjects are necessary. Therefore subjective evaluation methods for watermarked objects can take advantage of this procedures in order to gain quantative insights in the quality of the watermarked objects.

The two-alternative-forced-choice test If the impairments introduced by the coding procedure are very small one can assume transparency of the coded signal. To further validate this hypothesis, a subjective evaluation test for non-transparency can be performed by a so-called *two-alternative-forced-choice test*. In this case the hypothesis of non-transparency is tested (in contrast to the additional rating described below). A training phase precedes the actual test phase. During the training the test person compares the original and the watermarked item until he or she believes to perceive a difference. For the actual test, a number of pairs are randomly chosen from the set of possible combinations $\{(\mathbf{c_o}, \mathbf{c_o}), (\mathbf{c_o}, \mathbf{c_w}), (\mathbf{c_w}, \mathbf{c_o}), (\mathbf{c_w}, \mathbf{c_w})\}$, where $\mathbf{c_o}$ denotes the original and $\mathbf{c_w}$ the watermarked item. For each of these pairs, the subject is asked whether both items were equal or not. A correct decision about items being equal or different is called a "hit", so a subject produces a result of the form "k hits out of a number of n trials". During the test phase no limit is imposed on the number of repetitions to compare each of the individual items in the pair for comparison. Since the evaluation is performed for a group of persons, the hits within this group are summed and taken as the test variable.
A test for non-transparency is performed by trying to reject the transparency hypothesis. For this reason the following null and corresponding alternative hypotheses are formulated:

H_0 : Distortions are not perceivable.

H_1 : A subject can perceive distortions in a watermarked item.

What is tested is the ability to detect differences between the original and the watermarked object. The test variable is the number of hits k out of the number of pairs n. Under the hypothesis of H_0 the probability to get k hits out of n with detection probability $p = 0.5$ – because the subject is simply guessing – is:

$$P(k, n, p) = \binom{n}{k} p^k (1-p)^{n-k} = \binom{n}{k} 0.5^n \tag{7.1}$$

Therefore the distribution function of k is a binomial distribution $B(n, p)$. The critical region can be determined according to the following equation by choosing a level of significance α:

$$P(T \in \mathbf{B} \mid H_0) \leq \alpha \tag{7.2}$$

Choosing the level of significance $\alpha = 0.05$ and applying equation (7.2) for $n = 40$ pairs leads to a critical region of $B = \{26, \ldots, 40\}$, i.e. if a subject has more then 25 hits, the error probability of *wrongly rejecting* the null hypothesis is 5%.
To ensure independent experiments the best approach would be to use as many subjects as possible each one testing only one pair. If the subjects are able to distinguish between the original and the watermarked version they will do it with a certain probability of detection $p > 0.5$. In general we have no knowledge about this detection probability. The parameter $p = 0.5$ corresponds to the null hypothesis H_0 whereas all values of $0.5 < p \leq 1$ correspond to the alternative hypothesis H_1 of non-transparency.
What can be calculated is the so-called operation characteristic (OC) function $\beta(p)$. $\beta(p)$ is a function of the detection probability p and reports the error of *wrongly accepting* H_0 if the alternative hypothesis H_1 is correct:

$$\beta(p) = \begin{cases} 0.95, & p = 0.5 \\ P_p(T \notin \mathbf{B} \mid H_1), & 0.5 < p \leq 1 \end{cases} \tag{7.3}$$

The adjustment of the OC-function $\beta(p) = P_p(T \notin \mathbf{B} \mid H_1)$ is done according to equation (7.3). By using the OC-function we have the full knowledge about the errors and therefore the quality of the hypothesis test. The quality of the test strongly depends on the number of experiments performed. For example if the subjects are able to distinguish with a detection probability of $p = 0.7$ the error probability is

$$\beta(p) \approx \begin{cases} 0.617, N = 10 \\ 0.007, N = 100 \end{cases} \tag{7.4}$$

The quality of the test increases with a higher number of tested pairs (see figure 7.1). This can be used in order to design the test by

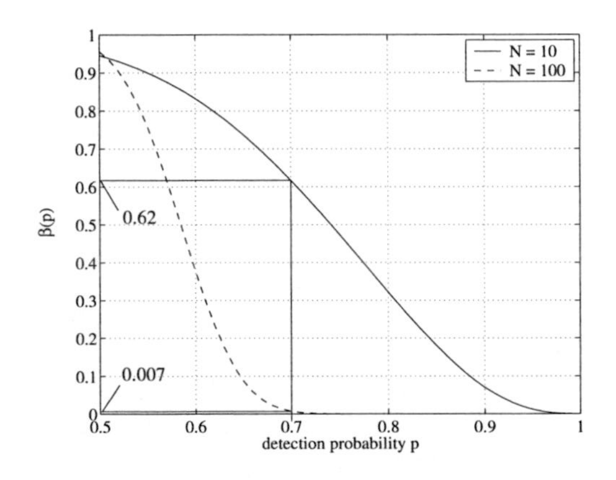

Figure 7.1: Quality function for hypothesis test

calculating the number of pairs needed to ensure the two kind of errors incorporated in hypothesis testing if the detection probability of the subjects are known.

7.2.2 The ITU-R BS.1116 Standard

The standard for subjective evaluations of small impairments of high quality perceptual audio coders is specified in the International Telecommunication Union (ITU-R) Recommodation BS.1116 [46] [2].This procedure is a so-called *double-blind A-B-C triple-stimulus* hidden reference comparison test. Stimuli A contains the reference

[2]It was published in 1994 and updated in 1997.

signal, whereas B and C are pseudorandomly selected from the coded and the reference signal. Therefore either B or C is the hidden reference stimulus. After listening to all three items, the subject has to decide between B and C as the hidden reference. The remaining signal is the suspected impaired stimulus. This one has to be graded relative to the reference signal by using the five-grade impairment scale according to ITU-R BS.562 [45].

Impairment	Grade	SDG
Imperceptible	5.0	0.0
Perceptible, but not annoying	4.0	-1.0
Slightly annoying	3.0	-2.0
Annoying	2.0	-3.0
Very annoying	1.0	-4.0

Table 7.1: ITU-R five-grade impairment scale

Table 7.1 contains an absolute and a difference grade. The "Grade" column can be treated as a continuous 41-point absolute category rating (ACR) impairment scale. It is used by the listener to grade the impaired signal relative to the reference signal. The stimulus which is identified by the subject as the hidden reference will be assigned the default grade of 5.0. The results of the listening tests are based on the so-called subjective difference grade (SDG) shown in the right column of table 7.1. It is calculated from the results of this rating by subtracting the score assigned to the actual hidden reference signal from the score assigned to the actual coded signal:

$$SDG = \text{Score}_{\text{Signal Under Test}} - \text{Score}_{\text{Reference Signal}} \tag{7.5}$$

Transparency is assumed if the SDG value is 0 whereas a value of -4.0 is very annoying.

Besides the fact that rigorous subjective listening procedures as described above are still the ultimate quality judgment they do have some disadvantages:

- The test results are influenced by the variability of the expert listeners. Experiments have shown that the various experts are sensitive to different artifacts [93], [83], [95].
- Playback level (SPL) and background noise can introduce undesired masking effects.
- The method of presenting the test items can have a strong influence on the quality (influence of loudspeakers and listening room of the specific site).
- Listening tests are time consuming.
- The equipment necessary to perform listening tests are very cost intensive.

Therefore the need of automatic perceptual measurement of compressed high-fidelity audio quality has motivated research into development of corresponding schemes. In the same way are objective measurements tools superior to subjective listening tests during the development phase of new audio watermarking algorithms, because of the lot of effort and time which has to be invested.

7.3 Objective Measurement of High Quality Audio

The ultimate goal of the objective measurement algorithms is to substitute the subjective listening tests by modeling the listening behavior of human beings. The output of the algorithms for objective measurements is a quality measure consisting of a single number to describe the audibility of the introduced distortions like the SDG in subjective listening tests. The various algorithms for objective measurement of audio quality fit into the principle architecture according to figure 7.2 on page 133. A difference measurement technique is used to compare the reference (original) signal and the test (processed, i.e. compressed or watermarked) signal. Both the reference

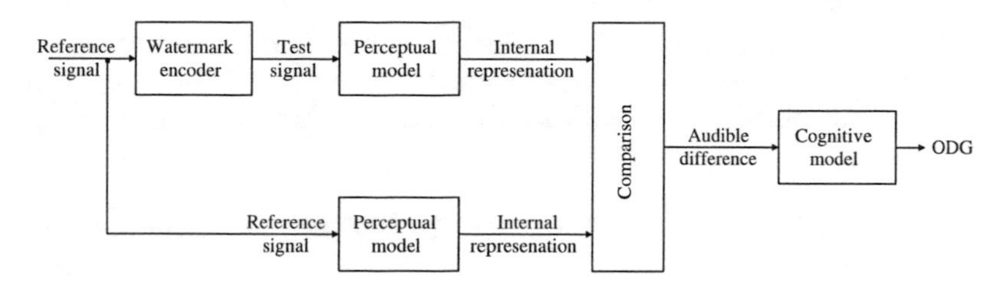

Figure 7.2: General architecture for objective quality measurement of audio data.

signal and the signal under test are processed by an ear model, which calculates an estimate for the audible signal components. This components can be regarded as the representation of the signals in the human auditory system. The *internal representation* is often related to the masked threshold, which in turn is based on psychoacoustic experiments performed by Zwicker [109]. From the two different internal representations of the reference and the test signal the *audible difference* is computed. Because the results of the listening tests are judged with a single SDG value, the corresponding measure has to be derived from the audible difference. This is accomplished by the *Cognitive Model*, which models the processing of the signals by the human brain during the listening tests. The output of the whole system is a the so-called Objective Difference Grade (ODG), which can be compared to the SDG in the listening test.
This has ultimatively led to the adoption of an international standard for the measure of the perceived audio quality (PEAQ), ITU-R BS.1387 [48]. The intention of this standard is to replace the described ITU-R BS.1116 standard, which is very sensitive and enables the detection of even small distortions. Since the intention of both the subjective and objective audio quality measurements is to compare the processed audio material with the original signal the test procedures are only useful in testing high quality audio. Applying these methods for the evaluation audio material with lower quality would lead to test results which were mainly directed towards the

bottom of the five-grade impairment scale (see table 7.1 on page 131) and therefore useless. For this reason the scope of the ITU-R BS.1116 and ITU-R BS.1387 standards are limited to data rates > 64kbit/s.

7.4 Testing Watermarked Items with Reduced Quality

In certain watermarking applications it might be reasonable to use data rates below the 64 kbit/s limit. In this case a problem arises with the evaluation of the quality by both subjective listening tests and objective measurement systems. As noted above the BS.1116 standard and its objective counterpart BS.1387 are not intended to use for bit rates < 64kbit/s. This limit might decrease in the future due to advances in the development of high quality perceptual audio compression coders. Furthermore new advanced subjective listening tests termed MUlti Stimulus with Hidden Reference Anchors (MUSHRA) are proposed by an European Broadcasting Union (EBU) project group [35]. Currently this test method is in the standardization process of the ITU-R [47]. In contrast to BS.1116 MUSHRA is a *double-blind multi stimulus* with hidden reference and hidden anchors. Since the subject will normally easily detect the larger distortions the usage of a hidden reference makes not sense in this test. The anchors are choosen according to the type of distortions the systems[3] under test are typically introduce.
The difficulty in evaluating of the impairments is the rating of the relative annoyances of the various artifacts. From this point of view the subjects have to decide if they prefer one type of impairment over the other. As a result comparison is not only made between the reference and system under test but with all other systems contributing to the test. A grading is performed between the different systems. It is derived by comparing that system to the reference signal as well as to the other signals in each trial. A trial in turn consists of the presentation of the reference signal, the anchors as well as all

[3]In the context of watermarking the term system is synonym to different watermarking systems.

versions of the test signal processed by the systems under test. In contrast to BS.1116 MUSHRA uses an absolute measure of the audio quality directly compared to the reference as well as the anchors. The grading scale in MUSHRA is the five-interval Continuous Quality Scale (CQS), which is divided into five intervals as shown in table 7.2.

Quality	Grade	Internal numerical representation
Excellent	5.0	100
Good	4.0	80
Fair	3.0	60
Poor	2.0	40
Bad	1.0	20

Table 7.2: Five-interval Continuous Quality Scale (CQS)

This absolute scale is necessary in order to be able to compare the results with similar tests. To summarize the different test methods one

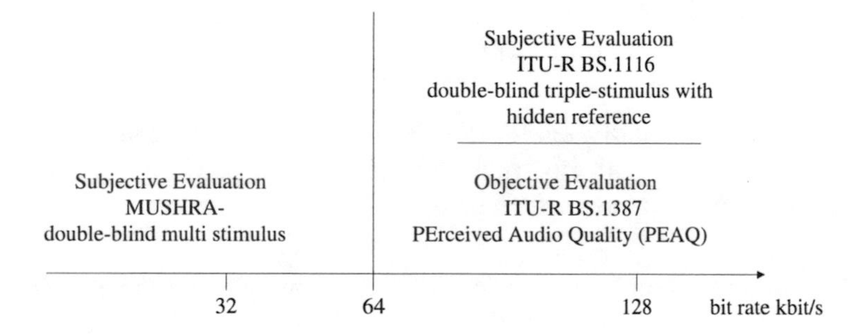

Figure 7.3: Quality measurement methods as a function of the bit rate.

has to consider the application and corresponding quality requirements, which should be specified in terms of bit rates. For example if the audio data are compressed during the application with a specific bit rate this figure can be used in order to decide which test method is appropriate for evaluating the quality of the watermarked items according to figure 7.3 on page 135.

7.5 Quality Tests and Results

7.5.1 Test Items, Parameters and Test Setup

To perform a reliable testing of the quality of watermarked audio tracks the whole test has to be specified in detail including:

- What is the aim of the test?
- What are the parameters used during testing?
- Explanation of the selection of test items used during testing.
- Description of the test equipment, setup and implementation.
- Statistical evaluation of the received test results in order to perform a quantative analysis.

Aim of this quality evaluation is to measure the transparency of the watermarked audio tracks with regard to different *NMR* (see equation (6.49) on page 116) settings.

Parameters The *NMR* has been used as the quality parameter [20] throughout the quality evaluation. The determination of suitable parameter values has been done by performing a subjective listening test as a function of the *NMR* (for three different settings):

- −5 dB (W_{Noise} 5 dB below the masking threshold)
- 0 dB (W_{Noise} exactly on the masking threshold)
- 2 dB (W_{Noise} 2 dB above the masking threshold)

Since the optimal *NMR* value may be dependent on the group of listeners and the test item, the test is performed for different items and evaluated with regard to different types of listeners. The different audio tracks are additional parameters used in the evaluation of the quality evaluation.

Test items used for this test are:

Harpsichord Testing hard attacks (origin SQAM [99] material).

Marissa Testing pop music.

Speech Testing German language (origin SQAM [99] material).

Vivaldi Testing classical orchestral music.

All audio tracks are excerpts from the original tracks with an amplitude resolution of 16 bits and a sampling rate of 44.1 kHz. The duration of each excerpt is about 10*s*.

Test setup A test is performed on each combination of the items and *NMR* setting. A training phase precedes the actual test phase. During the training the test person listens to the original and the watermarked item until he or she believes to hear a difference. The whole listening test was performed under the following conditions:

- The test person is alone in the listening room.
- The test is performed by using a high quality equipment:
  - Headphone amplifier STAX Vacuum Tube Output Driver Unit
  - Headphone STAX Signature Series II
  - Sound Card, SEK'D RME-Audio Digi 96
  - DA/AD Converter SEK'D ADDA 2496 S
- No listening time limit during the training and test phase.

- Full knowledge which item is presented during the training phase.
- The listener is guided through the test by a web application, which is also used to record the results.
- The results are stored in a database for the statistical evaluation.

For the actual test of one setting (item, *NMR*), ten pairs are randomly chosen from the four possible combinations $\{(\mathbf{c_o}, \mathbf{c_o}), (\mathbf{c_o}, \mathbf{c_w}), (\mathbf{c_w}, \mathbf{c_o}), (\mathbf{c_w}, \mathbf{c_w})\}$ and presented to each test person[4].

7.5.2 Test implementation

To simplify the test, a web-based application was created. The application runs on a server and generates HTML documents that are sent to the browser of the test person and make up the graphical user environment. The main reason for the decision of a web-based application is the great number of potential test persons reachable by the world wide web. Since the volume of data being transmitted for such a test is fairly large, we decided to present two alternatives for the test:

- Online evaluation
- Offline evaluation

The online evaluation is intended to be used if the test person has access via a high bandwidth connection to the server. This is especially the case when performing a local test. During the online evaluation, the subject is interactively led through the test by the application, presenting the different items and collecting the results from the user.
If the subject does not have a fast connection to the server, it still can participate via the offline evaluation. Therefore all pairs are

[4] $\mathbf{c_o}$ denotes the original and $\mathbf{c_w}$ the watermarked item as described in section 7.2.1.

randomly generated and put on an audio CD that can be listened to by the subject. After all items are compared, the subject fills in a HTML form also contained on the CD and sends the results back to the server to be stored in the database. This reduces the data being transmitted between server and browser because no audio files need to be loaded from the server. Furthermore the same code is used for the pseudorandomly selection of the pairs which are used in the offline or online evaluation. The selection is stored in the database for the comparison with the test results.

7.5.3 Subjective Listening Test Results

Three types of listeners where differentiated throughout the test presented in [9]:

Professional listeners E.g. sound engineers.

Musicians Persons playing any music instrument.

Normal listeners No special knowledge in music.

Since the evaluation is performed for a group of persons, the hits within this group are summed and taken as the test variable. The number of tries is the product of the number of persons within a group and the number of pairs for a combination of a specific item and *NMR* setting. Furthermore the evaluation can still be done for all subjects (without consideration of the group). The evaluation was performed with a total number of 20 persons. 16 persons belong to the group of *normal listeners*, four were *musicians*. The critical regions for each group are shown in the following table:

Listener group	Critical region B
Normal listeners	$\{91,\ldots,160\}$
Musicians	$\{26,\ldots,40\}$
Global	$\{112,\ldots,200\}$

Item	NMR/dB	Average Listener		Musicians		Global		Sound Engineers	
Speech	−5	82	yes	16	yes	98	yes	56	no
	0	92	no	29	no	121	no	75	no
	2	123	no	30	no	153	no	73	no
Marissa	−5	84	yes	19	yes	103	yes	40	yes
	0	87	yes	19	yes	106	yes	44	yes
	2	81	yes	23	yes	104	yes	40	yes
Harpsichord	−5	105	no	26	no	131	no	73	no
	0	142	no	37	no	179	no	79	no
	2	147	no	39	no	186	no	80	no
Vivaldi	−5	83	yes	18	yes	101	yes	44	yes
	0	75	yes	18	yes	93	yes	44	yes
	2	76	yes	22	yes	98	yes	50	no

Table 7.3: Result table for subjective listening tests

Two watermarks were simultaneously embedded to all audio tracks. The data rate for each watermark was $8bit/s$. The results of both tests are included in the table 7.3.

Combinations of items and *NMR* settings are marked with either "yes" or "no" depending on whether they are considered to be transparent or not. A second listener test was performed with *professional listeners* consisting of sound engineers from the Bertelsmann Group who tested the watermarked items in their studio. Therefore the results cannot be summed up with the previous one. The critical region for $n = 8$ professionals is $B = \{47, \ldots, 80\}$. The results of this test are reported in the last column of table 7.3. The evaluation shows the same result for the average listener and musicians. However the sound engineer are more critical as expected. The item *Harpsichord* cannot be considered transparent for any *NMR* setting. Also *Speech* is not transparent for the two worst quality settings for the average listener and the musician. Furthermore it is not transparent for the sound engineers in all *NMR* settings. These two items are the critical ones. Both tracks have a simpler structure than the *Vivaldi* and *Marissa* tracks, which of course makes it more difficult to embed the watermark without introducing audible distortions. The two remaining tracks *Vivaldi* and *Marissa* are more average material which is transparent for all groups of listeners in the two best quality settings.

7.5.4 Objective Measurement Results

This test presented in [3] rates the quality of the watermarked items with the PEAQ measurement system. The result of the test is the *ODG* parameter as a function of the test item and the *NMR* setting as can be seen from the following table 7.4:

Item	*NMR/dB*	PEAQ/*ODG*	Interpretation
Speech	−5	0.02	imperceptible
	0	-0.28	imperceptible
	2	-0.42	imperceptible
Marissa	−5	-0.04	imperceptible
	0	-0.37	imperceptible
	2	-0.70	imperceptible
Harpsichord	−5	-0.72	imperceptible
	0	-2.23	slightly annoying
	2	-2.64	annoying
Vivaldi	−5	-0.01	imperceptible
	0	-0.25	imperceptible
	2	-0.50	imperceptible

Table 7.4: Result table for objective measurement of the quality.

All but the *harpsichord* in the settings $NMR = 0, +2$ dB are graded in the imperceptible region $ODG = -0.5, \ldots, 0.0$.

7.5.5 Comparison of Subjective and Objective Test Results

The results of the subjective listening test according to table 7.3 on page 140 and the objective evaluation (see table 7.4) match for the

Marissa and *Vivaldi* audio tracks. The results for the *Speech* track do not match for the $NMR = 0, +2$ dB setting even though the results for the $NMR = 0$ dB are near the edge of the critical region. The same is true for the *Harpsichord* item in the setting $NMR = -5$ dB. The comparison of the results leads to the conclusion that the objective measurement is less critical in evaluating the quality of the watermarked tracks. This validates the fact that the two-alternative-forced-choice test is very sensitive. Furthermore since the computed *ODG* is an averaged value over the whole audio track the results have to be interpreted carefully. After a sufficient training phase a listener is capable to detect even the slightest impairments happening at a certain time frame. Such deviations are not seen anymore in a time averaged value like the *ODG* reported above.

7.5.6 Comparison of the Results of Objective Quality Rating

Contradicting requirements of the watermarking algorithm like *quality, robustness, security* and *application criteria* cannot in general be fulfilled simultaneously for all applications, as already discussed in section 6.2 on page 89. Due to the fact that a lot of audio material is distributed in a compressed format - which reduces the quality even further - it makes no sense to embed a watermark with a quality, which will be decreased in a stronger manner by a following compression process. In this case the application specific quality requirement can be formulated as a function of the bitrate of the compressed audio stream and the quality of the watermarked audio track respectively the *NMR* setting adjusted accordingly. To perform the comparison of the quality of the watermarked items with the audio tracks compressed with the MPEG 1 Layer III compression algorithms the audio tracks were compressed. Different bit rates were used in the compression of the items. The quality of the compressed test pieces were evaluated like the watermarked ones. The result of the test is the *ODG* parameter as a function of the test item and the bit rate setting as can be seen from the following table: Only for the highest bit rate of 128 kbit/s, which is judged as approximately CD quality, (see table 4.1 on page 55) the items are graded in the imperceptible region $ODG = -0.5, \ldots, 0.0$. As it can be seen

Item	*kbit/s*	PEAQ/*ODG*	Interpretation
S	128	-0.19	imperceptible
	96	-1.61	slightly annoying
	64	-3.46	very annoying
M	128	-0.74	imperceptible
	96	-2.39	slightly annoying
	64	-3.51	very annoying
H	128	-1.05	perceptible
	96	-3.42	very annoying
	64	-3.53	very annoying
V	128	-1.00	perceptible
	96	-2.13	slightly annoying
	64	-3.50	very annoying

Table 7.5: Result table for objective measurement of the quality of compressed items.

the PEAQ measurement for data rates of $64 kbit/s$ leads to test results which are concentrated at the bottom of the five-grade impairment scale (see table 7.2 on page 135). This was the reason for the development of the new advanced listening test MUSHRA, which can also be used to judge watermarked items with reduced quality as already mentioned in section 7.4. The results of the objective rating of the watermark according to table 7.4 can be compared to that of the perceptual audio codec according to table 7.5. The direct comparison of the results are shown in Fig. 7.4 on page 144. It can be seen that the watermarking introduces less distortions than the compression algorithm for all items for $NMR = -5$ dB. Furthermore the *Marissa* and *Vivaldi* track performs better than the lowest compression rate of 128 kbit/s for all settings of the *NMR*.

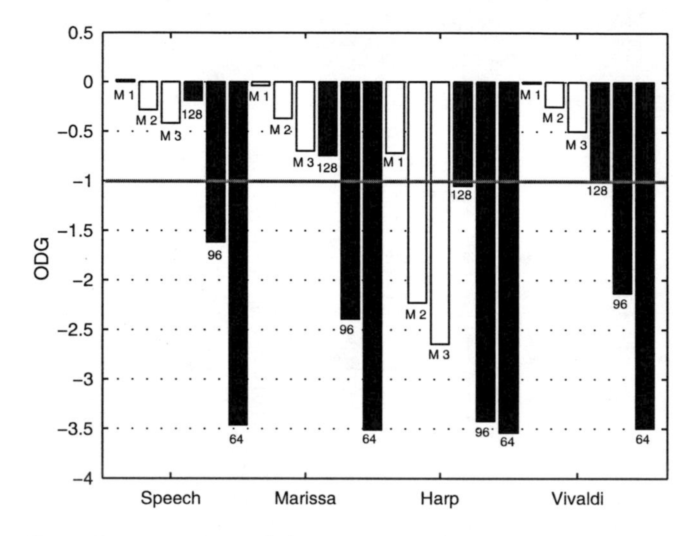

Figure 7.4: Comparison of the quality of watermarked items with MPEG 1 Layer III coded items.

7.6 Summary

This chapter described quality evaluation methods of watermarked audio tracks and performed the quality evaluation to the audio tracks watermarked with the audio watermarking method developed in chapter 6. The application of the procedures presented was motivated by the similar problem in the evaluation of the performance of compression coders with regard to the quality of lossy compressed audio data. In contrast to the compression coders, which add quantization noise during the encoding process the watermark embedder adds the watermark noise, which has to be kept inaudible. Both subjective listening tests and objective measurement methods were transferred to the problem of evaluating the quality of watermarked audio tracks with small impairments. Furthermore the problem of judging watermarked audio tracks with reduced quality by applying the so-called MUSHRA test was addressed. Both subjective listening tests namely the psychophysical 2AFC test and a objective measurement the PEAQ test were used to judge the quality of the audio tracks watermarked with A©WA. The results of the tests demonstrated the transparency of the watermarked audio tracks for average pop and classical pieces of music. This was verified by the objective measurement tests which correlate with the results obtained in the subjective listening tests. The comparison of the objective results of audio data compressed with MPEG 1 Layer III at different bit rates and the watermarked audio tracks verified that the quality of the watermarking items are significantly better than the corresponding compressed files.

CHAPTER

EIGHT

Benchmarking the A©WA Algorithm

As described in chapter 3, audio watermarking systems can be applied in a variety of areas. Depending on the use of the watermarking method, the security of the digital audio watermarks have to fulfill different requirements, some of which were also qualitatively mentioned in chapter 3. Therefore, according to the design of the underlying application, the identification of possible risks with respect to the usage of a watermarking algorithm is of particular importance. The risk analysis not only reveals security flaws of the application scenario as a whole but also determines the parameter settings of the used audio watermarking algorithm.

This chapter starts with an approach to identify security requirements in section 8.1. The security of the watermarking can be attacked in several ways. This not only depends on the effect the attack should produce but on the assumptions which can be made about an attacker in a specific application scenario. Therefore, currently known attacks and a classification thereof are the subject of section 8.2. Each class of attacks will then be explained in more detail in a corresponding subsection with an emphasis on audio data.

The classification of possible attacks enables the evaluation of the developed audio watermarking algorithm with regard to the different identified groups of attacks. Section 8.3 starts with a discussion presenting the benchmark parameters used in the evaluation procedure of the A©WA algorithm. The following subsections are presenting detailed test results about the robustness of the embedded digital audio watermarks and the security of the algorithm. Finally section 8.4 summarizes this chapter.

8.1 Threats and Risk Analysis

In each application requiring a certain level of security, the weakest point or points of the system is trivially where an adversary – assuming knowledge of the overall defensive system – will attempt to circumvent the implemented security mechanisms or elements lacking in such mechanisms. A first step in the identification of underlying risks and possible attacks, the application and the corresponding use cases of the watermarking system have to be defined very carefully. This includes the listing of the participating groups and their allowed permissions to perform certain operations.[1] Concerning watermarking the canonical operations are:

- embedding of watermarks
- detection of watermarks
- removal of watermarks

The removal of watermarks is always an unacceptable operation in security-related applications. Otherwise, if the watermark does not serve security-related issues, is irrelevant for an attacker – e.g. in case of annotation watermarks described in section 3.3. Nevertheless the watermark has to be robust against processing manipulations which can occur in the specific application.
The allowed operations applicable to the individual classes of actors can be collected in the so-called *operational table* (see table 8.1 on page 149).
Having specified the allowed permissions for each group in turn defines the requisite security properties of the used watermarking

[1]This approach is taken from [27].

Watermarking Operation	Remove	Embed	Detect
Publisher or Label	-	Yes	Yes
Radio station	No	No	No
Public	-	-	-

Table 8.1: Operational table for playlist generation

system. Given the possible attacks for each group identified by having an entry "No" in their row of the so-called operational table, an additional table can be created. The new table is an excerpt from the so-called attack table which lists the possible attacks according to the operation performed (this corresponds to the column of table 8.1) and the assumptions about the attacker (see section 8.2.1).

8.2 Attacks on Audio Watermarks and Countermeasures

An attack can be described as any processing which circumvents the intended purpose of the watermarking technique for a given application [4]. According to this definition, watermarking attacks include normal processing operations like lossy compression, D/A-A/D conversion, sample rate conversion, etc. which may happen in an application and unintentionally destroy the watermark. An attack potentially breaks the robustness of the watermark, which in turn is related to the quality of the *attacked* data. For this reason an attack is successful if it defeats the watermark technique while ensuring the quality according to the specified constraints of the application scenario.

8.2.1 Classification of Attacks

In order to easily identify the attacks, a classification of the attacks into several groups helps both the developers of watermarking al-

gorithms and users of the watermarking systems in identifying the security requirements as well as judging the usability of watermarking technology.
This may be of vital importance for an entire application scenario, since for some attacks no reliable countermeasures are presently known. Watermarking techniques can be foiled in several ways which are a direct consequence of the permitted operations of embedding, detecting and removing of the watermarks.
Embedding of the watermark always requires watermark detection. Three major categories of attacks rendering watermarking useless during detection can be identified:

- Watermarks cannot be detected. There are two strategies to obtain this result. To remove the watermark or to misalign the embedded watermark and corresponding detector.

- False watermarks are detected. This can be accomplished by attacks which perform some kind of embedding of false watermarks.

- Unauthorized detection of watermarks. Not carefully designed algorithms can produce false alarms.

The attack table (see table 8.2 on page 151) provides a broad overview over all the possible attacks – grouped according to the scope of the two parameters – results of the attack, and assumptions about the attacker.
Different types of attacks are possible depending on the knowledge of an attacker, the tools he has at his disposal, and on the availability of watermarked versions of the same or different works. Each row in table 8.2 corresponds to a different assumption about the attacker and represents a variation of one category. The three columns of the table represent the major classes of attacks.
Usually, the class of attacks which produce the "no detection" result are further subdivided [29], [40] in two classes according to the way the intended effect is achieved. *Removal attacks* erase the watermark

form the watermarked work without using the key for embedding the watermark. *Desynchronization attacks,*[2] misalign the watermark detector and the watermark without removing the watermark information.

Effect	No detection	False detection	Unauthorized detection
Operation	Remove Desynchronize	Embed	Detect
No knowledge	Signal processing Misalignement	Copy attack	-
Algorithm published	Specific attacks	Deadlock attack	-
Marked works	Collusion attacks	Copy attack	-
Detector	Oracle attacks	Copy attack	False alarms
Encoder and detector	Custom-tailored oracle attack	Overmarking	False alarms

Table 8.2: Table of attacks

In general, the power of the attacks is increased by the knowledge and tools the adversary has available. For example, if the attacker has no prior knowledge about the algorithm but an embedder and detector, he can check the effect of watermark removal with denoising and filtering tools with the detector by embedding own watermark with the aid of the embedder.
The attacks differ in the complexity of the operations involved and the effort the attacker has to perform. Clearly, attacks requiring no prior knowledge constitute the most general form; these are often based on common signal processing operations. Having access to watermarked copies of the same work with different watermarks or different works with the same watermarks offers the possibility to apply different kinds of so-called *collusion attacks*.
Knowledge of the underlying watermarking algorithm should always be assumed in cryptographic systems (Kerckhoff's principle [55]), and by extension for any security system. If a detector is available to the adversary, other types of attacks are possible. Depending

[2]Also called presentation [54] or detection-disabling [40] attacks.

on the values reported by the detector one can apply the *sensitivity analysis* or *gradient descent attacks* (see section 8.2.2). If both an embedder and detector are available even more sophisticated attacks like the *custom-tailored oracle attack* [77] can be applied. These attacks all belong to the removal attacks described in following section; the desynchronisation attack class including specific examples are subject of section 8.2.3.
An interesting approach to the circumvention of watermarking techniques is the class of embedding attacks producing a misinterpretation of the detection results (see section 8.2.4). Therefore, the term *interpretation attacks* is also used to denote this class [54]. Finally the class of detection attacks is presented in section 8.2.5.

8.2.2 Removal Attacks and Manipulations

The removal of watermarks represents the most obvious form of attacking a watermark. The restoring of the original sound track would be the extreme form of this kind of attack. If an attacker has no prior knowledge about an algorithm he can apply distortions since most watermarks are vulnerable. The removal of watermarks may also happen unintentionally due to operations during the preprocessing of the data in certain applications.

Signal processing operations Assuming a differential between the quality of material and capabilities available to a creator and to the eventual user of the material, it can typically be assumed that the watermarked audio track is processed in some way during the transmission from the watermark embedder to the watermark detector (see section 2.3.1 on page 28). Since processing of the data, particularly in the audio field, is widespread, the International Federation of Phonographic Industry (IFPI) has specified in considerable detail the robustness of an audio watermark [43].
Besides the requirements of preserving the quality and a bandwidth of 20bps of the embedded data channel, robustness of the watermark against a wide range of filtering and processing operations is formu-

lated as necessary features of the watermarking technology. An even more detailed example of the catalogue of requirements which have to be met by the watermarking method is the audio broadcast monitoring scenario discussed in section 3.5.1 as specified by the EBU.
Additionally, signal processing manipulations can be used in order to remove watermarks. Even users with no special knowledge in signal processing can apply these operations by the usage of common consumer-grade or free software products for audio editing to perform filtering, denoising, compression (MPEG), etc. operations automatically. This is even more critical if the procedure – such as the one detailed in [54] (in the case of images) for removing watermarks – is widely distributed.
Robustness against common signal operations such as the addition of noise or localized signal distortions is often achieved by using spread spectrum signaling techniques (see section 2.3.1 on page 28) in the design of watermarking algorithms. Spreading the watermark energy over a large spectrum minimizes the spectral density and can – without additional consideration of features of the signal to be marked – impose a boundary on the quality degradation of the watermarked object. Since a naive attacker has to add enough noise in order to destroy the watermark, this makes such trivial attacks impractical.
Without specific knowledge about the underlying watermarking algorithm, an attacker can apply noise removal techniques with the assumption that the added watermark is noise-like. In [96] it was shown that the Wiener filter is the optimal linear-filtering/noise-removal attack for specific watermarking systems if the added pattern is independent of the cover object, the work and the watermark are drawn from zero-mean normal distributions $N(0,\sigma)$, and linear correlation is used as detection statistic. Furthermore, Su and Girod [96] showed the countermeasure against this type of attack by shaping the power spectrum of the added watermark according to the power spectrum of the original work:

$$\phi_{ww}(\omega) = \frac{\sigma_w^2}{\sigma_c^2}\phi_{cc}(\omega) \tag{8.1}$$

In this case, the power spectrum of the watermark signal $\phi_{ww}(\omega)$ is a scaled version of the power spectrum $\phi_{cc}(\omega)$ of the carrier object, where the σ_w^2 and σ_c^2 represent the variances of the distributions from the watermark and the carrier signal. In turn, if the watermark is designed according to equation (8.1) its perceptual qualities are very close to the original signal, and it is difficult to estimate (or separate) the watermark from the carrier signal. This is also known as the *power-spectrum condition* (PSC). The adaptation required by this condition can be performed explicitly by estimating the power spectrum of the original object or implicitly by embedding the watermark in the compressed domain with automatic adaptation to the power spectrum [39].
Two special signal processing attacks were presented by Craver et al. [30] in the course of the Secure Digital Music Initiative (SDMI) challenge. They are discussed in the following within the framework set forth by the SDMI challenge.
Watermarking technology is one of the key components of the SDMI system [89] to protect the music. In September 2000, SDMI initiated a public challenge to test technologies proposed to be used by SDMI including four watermarking technologies (denoted by the letters A, B, C and F). For the watermark challenge SDMI provided three files[3]:

File 1 The original song.

File 2 File 1, watermarked.

File 3 Another watermarked song.

The challenge for the attacker was to produce a file from File 3, transparent in terms of quality but with an undetectable watermark. Furthermore, an oracle was provided by SDMI where the submissions were judged on success or failure with a binary decision variable.
In this sense, the attack mechanism cannot be considered a pure signal processing attack because of the additional on-line oracle and the access to at least one original and corresponding watermarked

[3]The files were 2 min long in CD-format (44.1 kHz sampling rate, 16 bit amplitude resolution).

file. Nevertheless the way of removing the watermark from the watermarked songs was done with signal processing methods without knowledge of the underlying watermarking methods. Having access to the original (file 1) offers the attacker the possibility to analyze the difference signal[4] between the watermarked (file 2) and the original signal.
In the attack on challenge B, the difference of the FFT magnitudes of both signals revealed two notches around 2.8 kHz and 3.5 kHz for different segments of the watermarked sample. The attack than filled those notches with random but bounded coefficient values.
In the attack on challenge C the difference signal revealed an narrow-band signal centered around 1.35 kHz. Two attacks were performed in challenge C. The first one shifted the pitch of the audio by about a quartertone (a kind of desynchronization attack, cf. next section) to move the bursts away from the center frequency. Another one applied a bandstop filter with the center frequency of 1.35 kHz.

Specific designed attacks Having knowledge about the underlying algorithm, in contrast to above, enables the attacker to design an attack specific for an algorithm or a class of algorithms by finding and exploiting their weaknesses. One of the easiest form of attacks is applying some kind of filtering (see above). A simple low-pass filter can be applied if it is known from the underlying technology that the watermark energy embeds a significant amount of energy in the high frequency range.
Publishing the watermarking algorithm by the developer is a general principle derived from cryptography and formulated by Auguste Kerckhoffs in 1883 [55]: The security of a cipher or other mechanism must rely solely on the secret key not on the secrecy of the algorithm. This in turn enables experts to examine and validate the techniques or to publish potential security flaws. An example of disregarding the Kerckhoffs principle is the SDMI attack on challenge A from Craver et al. [30]. During the analysis of the frequency response of the watermarking process, Craver et al. discovered that the underlying

[4]This can be computed in the time or frequency domain.

algorithm is a complex echo hiding (see section 5.3 on page 70) system including multiple time-varying echos. With a basic knowledge about the principles of the algorithm, a patent search revealed more technical detail about the pattern used to implement the multiple time-varying echos.
Moreover, the search provided the attacker with the probable identity of the company which developed the technology. This again showed the validity of Kerckhoffs desiderata as some of the main principles also in the development of reliable watermarking systems.
Another example of a specific watermarking attack against spread spectrum methods is described by Langelaar et al. [63]. The general idea is to estimate the watermark from the watermarked object $\mathbf{c_w}$ by non-linear filtering techniques. In this special case the attack was demonstrated for image watermarking techniques. Nevertheless the general principle can also be applied in the case of audio file with the usage of corresponding filtering techniques. The difference between the watermarked and filtered watermarked object

$$\mathbf{w}' = \mathbf{c_w} - \mathbf{c}'_\mathbf{w} \tag{8.2}$$

is a first approximation of the watermark. Before subtraction the estimated watermark it is weighted with an experimentally determined scaling factor γ to yield the final approximation $\widehat{\mathbf{w}}$ of the watermark (see figure 8.1).

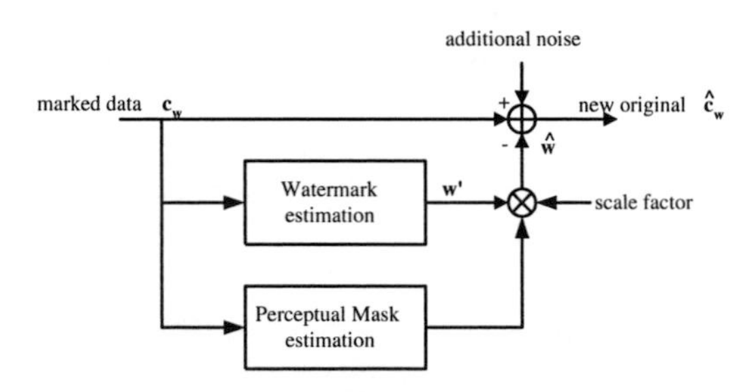

Figure 8.1: Removal by remodulation

Due to the use of estimation in finding the putative watermark signal, this class of attacks is also called *estimation-based attacks* [104]. In estimation-based attacks the knowledge about the watermarking technology and statistics of the original data and the watermark signal is taken into account as seen above.

Having estimated the watermark, an adversary may proceed in different ways and therefore the attack can be classified as removal, desynchronization or embedding attacks.

A good estimation of the watermark is necessary in two ways. First, the quality of the watermarked object is affected too much if a rough estimation of the watermark is substracted from the marked data (see figure 8.1 on page 156). Secondly, a rough estimation may not defeat the correlation-based detection of the embedded watermark.

In the context of this section *remodulation attacks* are a special form of estimation-based attacks trying to remove the watermark by performing a modulation inverse to the embedding of the watermark. The different blocks in the remodulation attack try to cope with the different requirements of quality of the watermarked object and the removal of the watermark. A scale factor is used to adjust between the distortions introduced in the watermarked object. A scale factor $\gamma > 1$ might reduce quality but may lead to a higher certainty of the removal of the watermark. An additional improvement of both goals of the attack can be achieved by calculating the perceptual masking threshold to weight the remodulated watermark. The basic assumption is that the perceptual masking threshold of the watermarked object is a good approximation of the masking threshold calculated from the original object. This assumption is valid if the embedding of the watermark to the cover object was parameterized in such a way as to be perceptually transparent.

In order to further decrease the performance of correlation-based detector, the attacker can add a significant amount of noise in less significant part of the data (see figure 8.1 on page 156). This is an approach which has been demonstrated for image data in [104].

Collusion attacks Even if the attacker has no special knowledge about the special algorithm or the class it belongs to, he can estimate the watermark or the original if he has more than one watermarked work. In this case the attacker can apply *collusion attacks*. Estimation of the watermark is possible if different works with the same watermark are available.
In the first case the attacker has access to $\{\mathbf{c_w}_i\}_{i=1}^n$ watermarked objects, all watermarked with the same watermark $\mathbf{w}$. He can obtain an approximation of the watermark by averaging of the watermarked works:[5]

$$\mathbf{w}' = \frac{1}{n}\sum_{i=1}^{n} \mathbf{c_w}_i \tag{8.3}$$

$$= \mathbf{w} + \frac{1}{n}\sum_{i=1}^{n} \mathbf{c_o}_i \tag{8.4}$$

This attack is possible, if the added watermark signal is not a function of the original work. Again, a possible countermeasure is to make the watermark dependent on the cover signal. An approximation of the original can be obtained if the attacker has the same work with different watermarks. In customer identification applications, where different customer IDs are embedded for identification purposes building a coalition between different customers can provide the access to the same watermarked piece with different watermarks. The same averaging process as described above can be performed which results in estimating the original cover signal:

$$\hat{\mathbf{c}}_\mathbf{o} = \frac{1}{n}\sum_{i=1}^{n} \mathbf{c_w}_i \tag{8.5}$$

$$= \mathbf{c_o} + \frac{1}{n}\sum_{i=1}^{n} \mathbf{w}_i \tag{8.6}$$

[5]This has e.g. been demonstrated for video applications [26].

A method to minimize this problem has been presented by Boneh and Shaw in the form of *collusion-secure codes* [17]. Boneh and Shaw showed that, if portions of the coded watermark are identical and carry enough information, at least one of the colluders can be identified. The proof relies on the assumptions that the identical parts of the coded watermarks is not affected by the above collusion attack.
A special form of collusion attack to recover the watermark is possible if it is embedded redundantly into distinct pieces of the carrier signal. In this case the carrier signal can be split in different pieces all containing the same watermark. One can regard the different pieces of one carrier signal as a set of different carrier signals, all containing the same watermark and apply the averaging described above.
This kind of attack was proposed by Boeuf and Stern [16] for audio watermarking technology F of the SDMI challenge (see above). Boeuf and Stern presented two ways to remove the mark from the file 2 in order to produce the unwatermarked file 3. The basic steps include the estimation of the watermark and a following removal from file 2.
By performing an autocorrelation of the difference between file 2 and file 1 (watermarked version - original), they found peaks with a constant distance of $l(\mathbf{c_w}_i) = 1470$ samples indicating the distribution of the same watermark over a number $N_c = \left\lfloor \frac{l(\mathbf{c_w})}{l(\mathbf{c_w}_i)} \right\rfloor$ of chunks $\mathbf{c_w}_i, i = 1, \ldots, N_c$ in the audio file $\mathbf{c_w}$ with $l(\mathbf{c_w})$ samples. A comparison of the individual chunks led to the insight that each of the chunks consisted of $N_{sc} = 10$ subchunks $\mathbf{c}_{\mathbf{w}.j}, j = 1, \ldots, N_{sc}$ (see figure 8.2) with the marked pattern $\mathbf{W} = (\mathbf{w}_1, \ldots, \mathbf{w}_{n_{sc}})$.

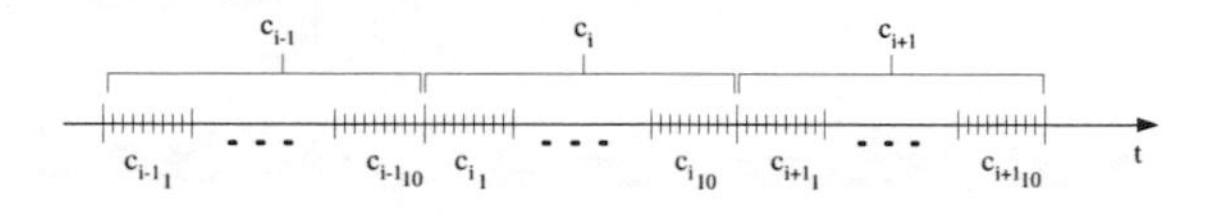

Figure 8.2: Segmentation in chunks and subchunks

In addition, the investigation by Boeuf and Stern showed that the watermark pattern was weighted by the product of the norm of each subchunk $\left\|\mathbf{c_w}_{i_j}\right\|$ and a multiplicative factor $\beta(\mathbf{c_o}, \mathbf{w}, i, j)$. Therefore the watermarked subchunk $\tilde{\mathbf{c}}_{\mathbf{w}i_j}$ was calculated according to equation (8.7).

$$\tilde{\mathbf{c}}_{\mathbf{w}i_j} = \mathbf{c_o}_{i_j} + \beta(\mathbf{c_o}, \mathbf{w}, i, j)\left\|\mathbf{c_o}_{i_j}\right\|\mathbf{w}_j, \quad i = 1, \ldots, n_c, j = 1, \ldots, N_{sc} \tag{8.7}$$

The norm of the signal $\left\|\mathbf{c_o}_{i_j}\right\|$ takes into account the power of the signal in subchunk j of chunk i, whereas the factor $\beta(\mathbf{c_o}, \mathbf{w}, i, j)$ is a slowly varying function which could not be fully determined. This knowledge was exploited during the recovery as well as the removal of the watermark. Two different kinds of watermark recovery were performed.
The first one estimated the watermark from the difference signal between marked and original version of file 1, whereas the more general method used only the watermarked file 2 without having access to the unmarked version of file 2. Both version had to renormalize the individual subchunks in order to eliminate influence of the normalization factor $\left\|\mathbf{c_o}_{i_j}\right\|$ in equation (8.7):

$$\mathbf{c'}_{\mathbf{w}i_j} = \frac{\tilde{\mathbf{c}}_{\mathbf{w}i_j} - \mathbf{c_o}_{i_j}}{\left\|\tilde{\mathbf{c}}_{\mathbf{w}i_j}\right\|} \quad \text{or} \quad \mathbf{c'}_{\mathbf{w}i_j} = \frac{\tilde{\mathbf{c}}_{\mathbf{w}i_j}}{\left\|\tilde{\mathbf{c}}_{\mathbf{w}i_j}\right\|} \quad \text{for} \quad i = 1, \ldots, N_c, j = 1, \ldots, N_{sc} \tag{8.8}$$

These normalized values were used in the averaging procedure, which was performed over all subchunks for a certain number p of chunks

$$\mathbf{w'}_j = \sum_{i=1}^{p} \mathbf{c'}_{\mathbf{w}i_j}, \quad p \leq N_c, j = 1, \ldots, N_{sc} \tag{8.9}$$

The actual removal of the watermark then consisted of the multiplication of the estimated watermark pattern $\mathbf{W'} = (\mathbf{w'}_1, \ldots, \mathbf{w'}_{N_{sc}})$ in each subchunk with the norm as weighting factor (see equation (8.7)) and the subtraction of this estimated signal from the watermarked signal (see figure 8.1 on page 156). Furthermore the attack by Boeuf and Stern demonstrated that the averaging attack is also possible if the added watermark signal is a function of the original signal as long as the function is known and can be approximated from the watermarked version.

Oracle attacks Even if the attacker has no knowledge about the algorithm or only one watermarked work he can apply *oracle attacks* if he has access to a watermark detector [67, 53]. This is the case in application scenarios where the attacker is allowed to detect watermarks, but not remove them, e.g. as in the SDMI scenario, which required the widespread distribution of watermark detectors [89]. The detector can be used as an oracle during attacking the watermark. Two kinds of oracle attacks relying only on the detector are possible, corresponding with the information which is returned from the detector.

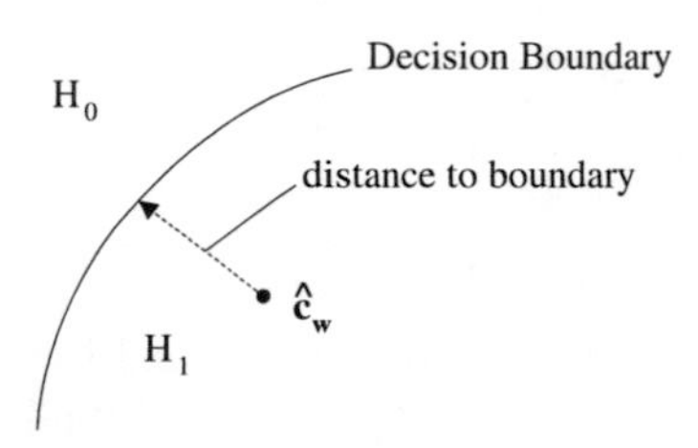

Figure 8.3: Decision boundary for binary hypothesis

A binary decision about the presence of the watermark is used in the *sensitivity analysis attack* (see figure 8.3), whereas the values of the detection statistic are exploited in the *gradient descent attack*. Both attacks use the detector response to find a short path out of the detection region. In the sensitivity analysis attack it is assumed that this path can be well approximated by the normal to the detection region.
In the gradient descent attack, the direction of the steepest descent is assumed to be a short path out of the detection region. This direction

is derived from the gradient of the detection statistic. Both attacks are performed in three steps:

1. Construct an object from the watermarked one, which lies near the detection region boundary. This can be done applying some kind of signal processing like filtering, compression etc. The thus constructed object can be a degraded version in comparison to the original. The construction is based on the response of the detector. If the output of the detector switches between "yes" and "no" in detecting the watermark, even on slightest modification of the altered object, it lies near the detection boundary.

2. Approximate the path (normal to the detection region or local gradient) out of the detection region. The approximation of the normal to the detection region is done iteratively, where the detection decision is recorded in each iteration. The local gradient is estimated by investigating the change of the detection value if the work is changed smoothly. The approximation of the normal [53] or the search of the local gradient can be implemented as an iterative process.

3. Scale and subtract the direction found in step 2, from the watermarked work (sensitivity analysis attack) or move the work along the direction (gradient descent attack) and repeat step 2.

A special form of an oracle attack called *custom-tailored oracle attack* can be performed if the attacker has access to the embedder and detector [77], i.e. in an extension to the oracle attacks discussed above. In this case the attacker embeds own watermarks with the embedder and removes the markings using the oracle attacks described above. The sensitivity-analysis attack relies on the assumption that the decision boundary of the binary hypothesis test can be estimated. Performing slight changes until the detector cannot detect the watermark any longer a large number of times yields different points of the decision boundary (see figure 8.4). This in turn is used to approximate the normal to boundary in order to find the minimum

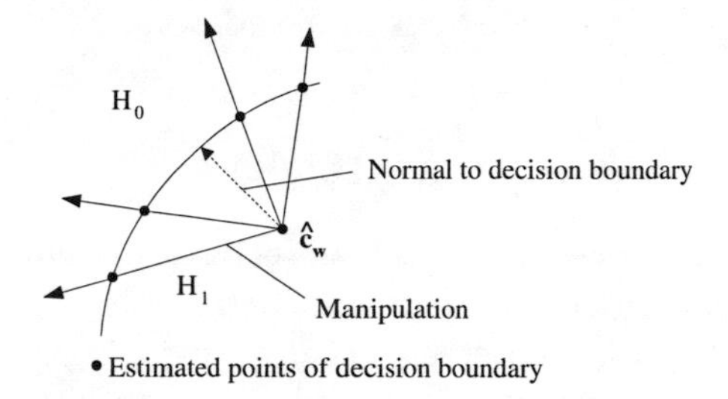

Figure 8.4: Approximation of the decision boundary

distortion path out of the detection region. This points to a possible countermeasure, making the decision boundary incalculable. Tewfik and Mansour present an approach where the decision boundary is modified to have a fractal shape [101].

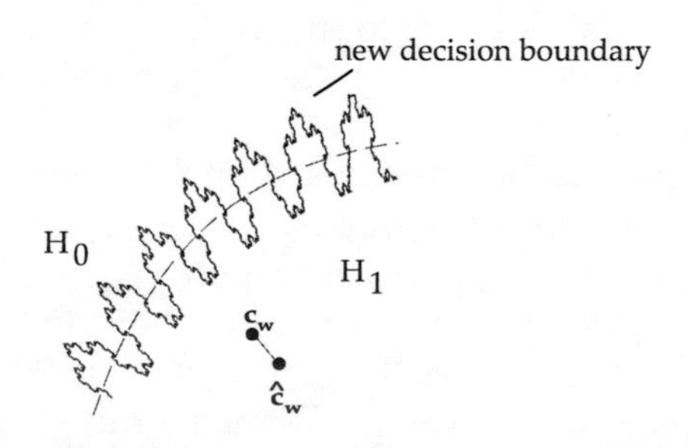

Figure 8.5: Modification of decision boundary

To retain the robustness of the watermark, the distance to the new fractal decision boundary is kept constant by modifying the watermarked object accordingly, which may introduce additional distortions and artifacts (see figure 8.5). As a result, the modification of the decision boundary has to be adjusted between the two conflicting needs that the decision boundary cannot be approximated even if the attacker has unlimited access to the detector and the requirement of preserving the quality of the watermarked object.

8.2.3 Desynchronization Attacks

The aim of the desynchronization[6] as well as the removal attacks is to render the embedded watermark undetectable. Nevertheless, the process of preventing the detection by means of desynchronization attacks is different. Instead of erasing the watermark, desynchronization misaligns the embedded watermark and corresponding detector in such a way that it is computationally infeasible to perform synchronization prior to detection.

Global and local transformations Most of the watermarking algorithms, especially those based on correlation, require perfect or near-perfect alignment during detection. Therefore applying global and local transformations aims at the destruction of the synchronization between the watermark and the detector. Global distortions of watermarked creations include delay or time-scaling for audio creations. More complex operations are pitch-preserving time-scaling and sample removal in audio tracks.
Breaking an audio watermarking technology by applying pitch shifting was already presented as a signal processing operation (cf. the discussion of SDMI attack challenge C in section 8.2.2). An example of a time-warping attack was performed in the SDMI attack challenge F by Craver et al. [30]. They warped the time axis by inserting a periodically varying delay defined by function $f(t)$

$$t'(t) = t + f(t) \tag{8.10}$$
$$\hat{c}_w(t) = c_w(t'(t)) \tag{8.11}$$

where $c_w(t)$ represents the marked signal and $\hat{c}_w(t)$ the attacked one.

[6] Also called misalignment attacks [27].

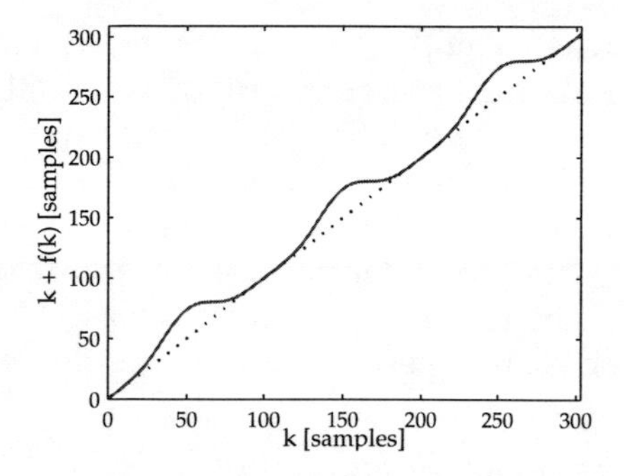

Figure 8.6: Example of the time warp with $f_s = 100, p = 2, s = 600$.

With the time $t[seconds]$ and the sampling rate $f_s \left[\frac{samples}{seconds}\right]$ the discrete-time function $f(k), k = tf_s$ is defined[7] by (see figure 8.6):

$$f(k) \;=\; \frac{d}{f_s}\left(1 - \cos\left(\frac{p\pi k}{f_s}\right)\right)^2, k = 0, \ldots, l(c_w) \qquad (8.12)$$

The number of samples for one period of the function $f(t)$ can be adjusted by p:

$$k = \frac{2}{p} f_s \qquad (8.13)$$

whereas the time delay per period is determined by the maximum number of samples $4d$ delayed:

$$\Delta t = \frac{4d}{f_s} \qquad (8.14)$$

The parameters during the SDMI attack $f_s = 44.1kHz, d = 6.75$ and $p = 0.602$ were derived from the study of SDMI challenge A. Therefore in a period $T = \frac{2}{p} \approx 3.32$ s the audio files were distorted by 27 samples or $\Delta t \approx 0.6$ ms.

[7]The choice of parameters are only for illustration purposes.

Scrambling attacks Another kind of desynchronisation can be performed by scrambling of samples or pieces of the watermarked creation prior to the presentation[8] to a watermark detector. If the watermarked creations are not directly modified but only their presentation the attacks are performed on a system level[9] which cannot be addressed within the watermarking system itself. Examples in the case of images include pixel permutations, the Mosaic attack [78] and the use of scrambling and descrambling devices in the case of video.

An advanced form of attack for any watermarking technology independent of the content is not to attack the watermarked version but to create a new work. The assumptions about the attacker are very low, since he/she does not need to know the algorithm or have access to the detector nor to more than one watermarked work. This kind of generic attack is simulated by the *Blind Pattern Matching* (BPM) *attack* [79]. The re-recording of the creation is done by replacing small pieces of the watermarked creation with perceptually similar pieces from the same watermarked creation or from an external library. A necessary assumption for a success of this attack is that the pieces used for substitution are containing different watermarks with little correlation to the original watermark. The assumption is valid even if the same watermark is embedded redundantly in a creation. Because this is often done in a block-based manner the probability that the substitution pieces are carrying the same part of the watermark is low if the size of pieces are one order of magnitude smaller than the size of the watermark. The attack was demonstrated by Petitcolas and Kirovski using the same watermarked audio file and permuting small pieces of it[10].

A possible countermeasure against this attack is to identify blocks with no similar counterparts and use only these blocks for embedding.

---

[8]These are also referred to as *presentation attacks*.

[9]This is the reason why they are often called *system level attack*.

[10]This explains the classification of BPM as scrambling attack.

8.2.4 Embedding Attacks

Embedding or ambiguity attacks simulate an embedded watermark even if it is not. The effect of this attack is the 'false detection' of watermarks in contrast to 'no detection' of the removal or desynchronisation attacks. Three main variations of this attack are considered in the next paragraphs according to the assumption which can be made about the attacker.

Copy attack The aim of the *copy attack* is to copy a watermark from one carrier signal to another. This attack is basically performed in two steps. In the first step an estimation of the watermark from the marked carrier signal is calculated. In the second step the estimated watermark signal is copied form the marked signal to the target carrier signal data to obtain a watermarked version (see figure 8.7). The

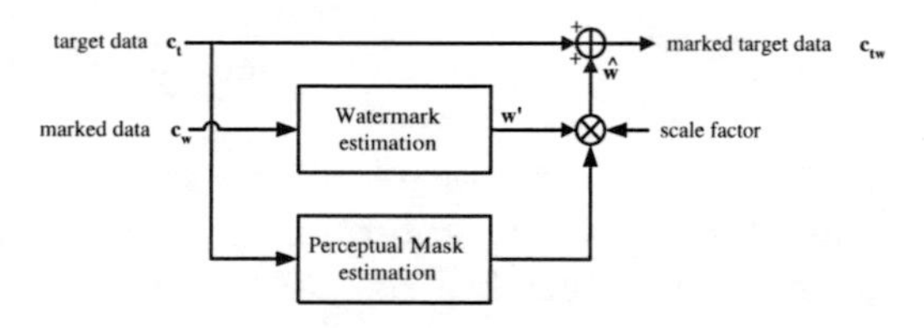

Figure 8.7: The copy attack

estimated watermark can be obtained in different ways depending on the assumptions made about the attacker. If the attacker has no prior knowledge about the algorithm but the same object carrying different watermarks he can perform the collusion attack (see section 8.2.2). This approximates the original object. The watermark he wants to estimate is then obtained by substracting the estimated original from the corresponding watermarked version (see equation (8.2) on page 156).
The copy attack was demonstrated for images by Kutter et al. [61], who performed a watermark removal attack to obtain the original via

spatial domain filtering. A previous approximation of the original is not necessary if the attacker can estimate the watermark directly. This is possible if he uses the first version of the collusion attack were access to different objects with the same watermark is required (see section 8.2.2).
A possible countermeasure to prevent the copy attack is to establish a link between the watermark and the carrier signal via cryptographic hash functions [27]. This link can be verified during the detection of the watermark. Another possibility maybe to make the watermark a function of the original carrier signal. In this case copying will be more problematic in terms of quality of the marked target carrier signal.

Overmarking Overmarking is an operation where a second watermark is embedded in an already marked carrier signal. Both watermarks can be detected independently if, for example, the locations where the information will be embedded is determined by the secret key.
This operation can always be performed if the attacker has access to the embedder and detector of the watermarking system. If the intention of the watermark is copyright protection, both parties the copyright owner and the attacker may claim ownership. The problem of ownership would be solved in this case if the order of watermark insertion can be proven reliably.
The only advantage the copyright owner has compared to the attacker is access to the true original. Since the attacker has only access to the already watermarked object, the sequence of embedding would be determined by the fact that both parties have to read the watermark from their corresponding putative original. A problem therefore arises if both parties can read their watermarks from the original of the opponent. In this case a stand-off is created, where the copyright owner has no real advantage over the adversary. This is aim of the so-called *deadlock* or *IBM attack*[11].

[11] Occasionally also called Craver attack.

Deadlock attack Different forms of the *deadlock attack* are possible dependent to the access of the attacker to the original creation or not. Let $\mathbf{c_o}$ be the original creation and $\mathbf{c_w}$ the watermarked version. To distinguish between the copyright owner and the attacker, the letters **c** and **a** will be used. The basic assumption of this attack is that the correlation between true watermark and the fake watermark is very low (this is very likely).

$$C_\tau(\mathbf{w_c}, \mathbf{w_a}) \approx 0 \tag{8.15}$$

Furthermore, the watermarked creation $\mathbf{c_w}$ and the fraudulent original are created according to the following equations:

$$\mathbf{c_w} = E_{K^c}(\mathbf{c_o}, \mathbf{w_c}) = \mathbf{c_o} + \mathbf{w_c} \tag{8.16}$$

$$\mathbf{c_a} = E_{K^a}^{-1}(\mathbf{c_w}, \mathbf{w_a}) = \mathbf{c_w} - \mathbf{w_a} \tag{8.17}$$

$$\Rightarrow \mathbf{c_w} = E_{K^a}(\mathbf{c_a}, \mathbf{w_a}) \tag{8.18}$$

$\mathbf{c_w}$ is constructed via an ordinary embedding process whereas the creation of $\mathbf{c_a}$ is based on the inversion of the embedding process of a watermarking system. It appears like a regular embedding function, which embeds the fraudulent watermark $\mathbf{w_a}$ into the fraudulent original $\mathbf{c_a}$ to yield $\mathbf{c_w}$.
Craver et al. [28] demonstrated this attack in a informed detection system by using the Cox algorithm [25] where the strength of the detection measured by C_τ were nearly identical. Using blind detection, the additional requirement that the correlation between the original and assumed watermark is zero (when compared to the threshold τ according to equation (2.5)) has to be made:

$$C_\tau(\mathbf{c_o}, \mathbf{w_c}) = C_\tau(\mathbf{c_a}, \mathbf{w_a}) = 0 \tag{8.19}$$

Disregarding this assumption would accept the existence of false alarms due to the correlation of the original creation and pseudorandomly generated watermarks. During blind detection Alice has to demonstrate that her watermark is embedded in $\mathbf{c_w}$ and the fraudulent original $\mathbf{c_a}$:

$$C_\tau(\mathbf{c_w}, \mathbf{w_c}) = \underbrace{C_\tau(\mathbf{c_o}, \mathbf{w_c})}_{=0} + C_\tau(\mathbf{w_c}, \mathbf{w_c}) = 1 \quad (8.20)$$

$$C_\tau(\mathbf{c_a}, \mathbf{w_c}) = C_\tau(\mathbf{c_w}, \mathbf{w_c}) - \underbrace{C_\tau(\mathbf{w_a}, \mathbf{w_c})}_{=0} = 1 \quad (8.21)$$

Again, Bob can also prove that his watermark is embedded in $\mathbf{c_w}$ and the original $\mathbf{c_o}$:

$$C_\tau(\mathbf{c_w}, \mathbf{w_a}) = \underbrace{C_\tau(\mathbf{c_a}, \mathbf{w_a})}_{=0} + C_\tau(\mathbf{w_a}, \mathbf{w_a}) = 1 \quad (8.22)$$

$$C_\tau(\mathbf{c_o}, \mathbf{w_a}) = C_\tau(\mathbf{c_w}, \mathbf{w_a}) - \underbrace{C_\tau(\mathbf{w_c}, \mathbf{w_a})}_{=0} = 1 \quad (8.23)$$

Therefore, in both detection cases the copyright owner would have no advantage in comparison to the attacker. This is even more surprising since the attacker has no access to the true original creation $\mathbf{c_o}$. The vulnerability exploited in producing this ambiguity is the invertibility of the watermarking algorithm according to equation (8.17). One approach is to make the watermark a function of the original, such that the fraudulent watermark cannot be created without access to the original.

$$\mathbf{c_a} \neq E_{K^a}^{-1}(\mathbf{c_w}, \mathbf{w_a}(\mathbf{c_a})) = \mathbf{c_w} - \mathbf{w_a}(\mathbf{c_a}) \quad (8.24)$$

According to equation (8.24) construction of the fraudulent original is not possible because the creation depends on the fraudulent original itself. One example of the function $\mathbf{w_c} = f(\mathbf{c_o})$ is to use the hash $H(\mathbf{c_o})$ over the original as the seed of the PN generator.
Another possibility is to use the hash $\mathbf{w_c} = H(\mathbf{c_o})$ as the watermark to be embedded. This is possible for every watermarking algorithm since the problem of preventing the inversion of the embedding function $E_K(\mathbf{c_o}, \mathbf{w})$ is shifted to the problem of inverting the function

$f(\mathbf{c_o})$ which is assumed computationally infeasible when using a cryptographic hash function.

8.2.5 Detection Attacks

The opposite of unauthorized embedding is unauthorized detection. Usually unauthorized detection of the embedded watermark is used as the preceding step before the corresponding removal. On the other hand *detection attacks* simulate the detection of watermarks even if these watermarks were not inserted afore. In this case the attack produces *false alarms*.

False alarm attacks Usually the effect of producing false alarms requires that the attacker has access to the detector. The question to ask is which kind of false alarm should be generated? Thinking about the copyright protection application the following question has to be answered. Is the watermark $\mathbf{w}$ embedded in the dataset $\mathbf{c_o}$? In this case applying a false alarm attack would require from the attacker to satisfy the following equations:

$$D_K(\mathbf{c_o}, \mathbf{w}) = \mathbf{w}' \quad \text{and} \quad C_\tau(\mathbf{w}', \mathbf{w}) \geq \tau \tag{8.25}$$

If the watermark is fixed the attacker has the possibility to vary the parameters $K, \mathbf{c_o}$. Varying the key K is equivalent to a search of a key for a fixed object $\mathbf{c_o}$. Using a fixed key would require the attacker to change the object until the equation (8.25) is satisfied. Having the detector in its hands this can be accomplished by performing the sensitivity analysis attack described in section 8.2.2. Of course both kind of approaches should in general be computationally infeasible for an watermarking algorithm.
Considering the number of attacks presented in the last section it seems to be a rather difficult task for the designer of a watermarking algorithm to cope with all or even a subset of possible attacks. Nevertheless, not every attack makes sense in an application and the applicability of the whole watermarking technology is related to the

quality of the data respectively the quality of the attacked data as presented in chapter 7.

8.3 Evaluating the A©WA Algorithm

Watermarking algorithms are mainly judged by two evaluation criteria. Their ability to preserve the quality of the original carrier signal and the robustness of the embedded watermarks. Subjective and objective methods for quality evaluation of watermarked audio tracks were presented in sections 7.2 - 7.4. The quality of audio tracks watermarked with the A©WA algorithm were evaluated in detail in section 7.5 as a function of the *noise-to-mask ratio* (NMR).

8.3.1 Evaluation Criteria and Metrics

However, besides the NMR serving as quality parameter further evaluation criteria might be relevant depending on the application. The concept of robustness is obviously and strongly connected to the quality of the watermarked items:

Definition 8.1 *Watermarks are robust if they cannot be destroyed without affecting the quality of the watermarked object in such a way that it is useless for the application.*

As already discussed in chapter 2, both criteria cannot be maximized simultaneously. It is impossible to ensure the highest quality of the watermarked signal and the maximum robustness of the embedded watermarks at the same time. Therefore, as discussed in chapter 3, evaluation of the usefulness of specific watermarking methods is always to be performed for a specific application and its requirements. A metric for measuring the robustness of the watermarks is the *bit-error rate* (*BER*), also known as the bit error probability. It is defined by the probability that binary data carried by the decoded watermark W' is different from the encoded W. Assuming N bits per watermark with ΔN different bits the bit-error rate is defined by

$$BER = \frac{\Delta N}{N} \tag{8.26}$$

To obtain reliable values for the BER, measurements have to be performed for different audio tracks, watermarks and keys. Besides the NMR specifying the quality on the one hand and the watermark strength on the other hand the robustness also depends on several additional parameters which are used as evaluation criteria.
Capacity is a general term for the number of bits which can be embedded in a certain media type that is occasionally found in literature. The more precise term for streaming data like audio is the *data rate*, which refers to the number of bits which can be embedded in a temporal segment of the carrier signal (usually one second). Accordingly, the data rate is measured in the number of bits per second (*bps*). This parameter is further related to the *watermark minimum segment* (WMS) specifying the minimum time segment of the carrier signal necessary in which a watermark can be embedded and reliable payload reliable extracted.[12]
Additionally *security parameters* not strictly related to the robustness of the watermarks have to be taken into account like the available key space (the total number of keys that can be used for embedding). Other parameters like the individual implementation decides about the security of the watermarking algorithm. Using a "bad" PRNG for the derivation of keying material will influence the security of the watermarking system, as will the implementation (e.g. tamper resistance of embedding and retrieval circuits as well as tamper resistance of the overall system – a defect in which may result in the elimination of watermark processing in its entirety from a protection system).
Furthermore the *complexity of the algorithm* is of vital importance in applications requiring the embedding of watermarks prior to the streaming or broadcasting of the audio material. Necessary preconditions of such application scenarios are the embedding and detec-

[12]This parameter is sometimes called granularity.

tion of the watermarks in real-time or even a multiple thereof (see sections 3.4 and 3.5).

8.3.2 Performance Evaluation and Representation

The amount of embedded information and the WMS are usually fixed in an application. In turn the data rate is determined by WMS and the algorithm in order to ensure the proper detection of the watermark according to the specified limits. Besides the fact that the available key space as well as the complexity of the algorithm are important parameters to characterize a watermarking algorithm they do not have a direct impact on the robustness of the watermark. The degree of reliance of the retrieving process characterizing the presence or absence of the watermark can be measured by the detection P_D and false detection P_F[13] probabilities.

P_D represents a hit, the probability that the watermark is indeed embedded and above a threshold defined by the algorithm. P_F is the probability for obtaining a hit despite there being no watermark signal present. The calculation of these probabilities require the knowledge of the underlying probability distributions of the original and marked version of the audio file. Because the distribution of the marked file are in general unknown, especially if the audio file experienced some sort of attacks, the P_D can only be determined experimentally.

Similarly the probability P_F that the retrieved watermark is a false one can be calculated from the experimentally obtained BER, which represents the probability that a single bit is flipped. The probability P_F that one or more bits are flipped in the retrieved watermark (consisting of n bits) can be calculated via

$$P_F(BER) = \sum_{k=1}^{n} \binom{n}{k} BER^k (1 - BER)^{n-k}$$

[13] Also called the detection-error.

$$
\begin{aligned}
&= \sum_{k=0}^{n} \binom{n}{k} BER^k (1 - BER)^{n-k} - (1 - BER)^n \\
&= 1 - (1 - BER)^n
\end{aligned} \tag{8.27}
$$

The factor $(1 - BER)^n$ in equation (8.27) represents the probability that the n bits of the watermark are retrieved correctly. P_F can be calculated precisely if the probability distribution of the original audio file is known. In this case BER_i represents the probability that bit i is flipped. In turn the probability for a correct bit i can be calculated as

$$1 - BER_i = \int_{-\infty}^{z_i} \Phi(t)dt \tag{8.28}$$

with $\Phi(z)$ the probability distribution of the unmarked audio track. This leads to

$$P_F = 1 - \prod_{i=1}^{n} (1 - BER_i) \tag{8.29}$$

Assuming a Gaussian distribution with standard deviation σ in the unmarked case

$$\Phi(z) = \frac{1}{\sqrt{2\pi}\sigma} e^{-\frac{z^2}{2\sigma^2}} \tag{8.30}$$

P_F can be calculated with

$$P_F = 1 - \frac{1}{2^n} \prod_{i=1}^{n} \left(1 + \operatorname{erf}\left(\frac{z_i}{\sqrt{2}\sigma}\right)\right) \tag{8.31}$$

using the definition of the error function [84]

$$\operatorname{erf}(x) = \frac{2}{\sqrt{\pi}} \int_0^x e^{-t^2} dt \tag{8.32}$$

In order to judge the robustness of the embedded watermark against an attack the BER can be plotted as a function of the strength of the attack, which is often determined by a parameter in the case of removal and synchronization attacks. Parameters of the watermarking algorithm like data rate and especially the quality have to be fixed. Usually the quality adjustments of the watermarked audio tracks have to be made according to the intended application.

8.3.3 Robustness against Removal Attacks

Not every attack or signal manipulation makes sense in every application as already mentioned in section 8.2.1. The signal manipulations performed to test the robustness of the watermarks are derived from the requirements formulated by the EBU (see table 8.3) for the application of the watermarking method in a monitoring scenario. All the audio files used during the test had the same standard CD

Manipulation	*Parameter*
ISO/MPEG 1 Layer III compression	192, 128, 96, 64 KBit/s
Re-sampling D/A → A/D conversion	
Stereo to mono conversion	
Sample frequency conversion	44.1 → 32 → 22.05 kHz
General multi-band equalizer	-6, . . ., +6 dB
Collusion and collusion-like attack	10 watermarked copies

Table 8.3: Signal manipulations performed during the removal tests

format, which consists of a stereo signal, sampled at 44.1 kHz having a amplitude resolution of 16 bit. In order to perform a reliable evaluation different host data, watermarks and keys have to be used. The watermarking parameter settings are the default settings concerning quality, data rate and hierarchy level (see table 8.4 on page 177). The different combinations of the number of host data, watermarks and keys correspond to a total number of 4000 watermarked audio tracks. One of the most important signal manipulations is the compression of the audio track via the ISO/MPEG 1 Layer III algorithm[14]. According to table 8.3 four different settings for the bit rate were used in

[14]Also known as MP3 algorithm.

Parameter	*Values*
Number of audio files	20
Duration	20 seconds
Watermark	10 different
Key	20 different
Quality setting	0 dB
Data rate	8 bit/second
Hierarchy level	1

Table 8.4: Parameter settings used during the removal attack evaluation

order to test the robustness against compression. The results shown in (see figure 8.8) verify the good robustness against MPEG compression down to 128 KBit/s. A significant increase in the BER can be seen if the bitrate is equal and below 64 KBit/s, which is not used for the typical distributed audio files on the internet, because of the low quality.

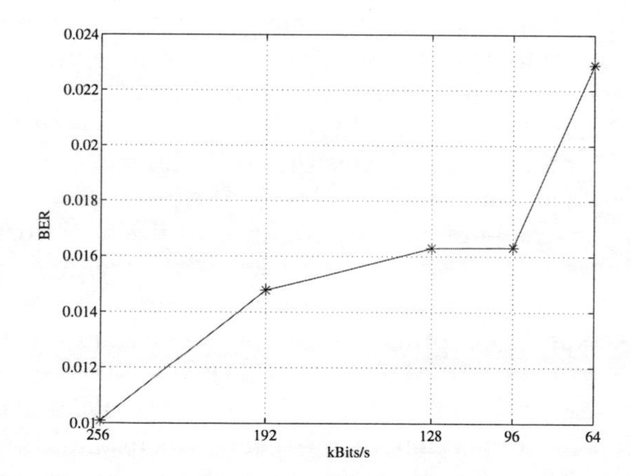

Figure 8.8: BER of watermarked and compressed audio stream

D/A → A/D was performed by re-recording the played sound files with a second soundcard. Due to the costly evaluation of recording each of the 4000 marked files, this test was performed for each of the files with one watermark and the different key combinations. Stereo to mono conversion was performed on all the marked files. The influence of this manipulation in comparison to the marked files before the conversion can be seen in in table 8.5.

Mono and sample rate conversion	*BER*
Stereo → Mono	0.0152
D/A → A/D	0.01001
Sample rate 44.1 → 32 kHz	0.01009649
Sample rate 44.1 → 22.05 kHz	0.01011257

Table 8.5: Mono and sample rate conversion and the corresponding BER

The results verify the good robustness against these signal manipulations. Sample frequency conversion is performed if audio files are combined with video for multimedia presentations. Furthermore sample rate conversion also come along during the compression of the audio files at lower bit rates. The robustness against this type of manipulation was tested by reducing the original sample rate of 44.1 down to 32 respectively 22.05 kHz (see table 8.5). A general multiband equalizer is often used to weight the frequency band differently to compensate for deficiencies in the recording or to generate a different sound impression. The robustness of the watermark against this signal processing was evaluated with the varying amplification parameter according to table 8.3 on page 176. The results of the watermarked and manipulated files are shown in figure 8.9.
Collusion and collusion-like attack are special kind of removal attacks as presented in section 8.2.2. Different kind of collusion attacks can be performed dependent on the result which should be obtained. The intention of the first variation is to estimate the watermark (see equation (8.3) on page 158). This kind of attack uses different audio files all marked with the same watermark using the same secret key.

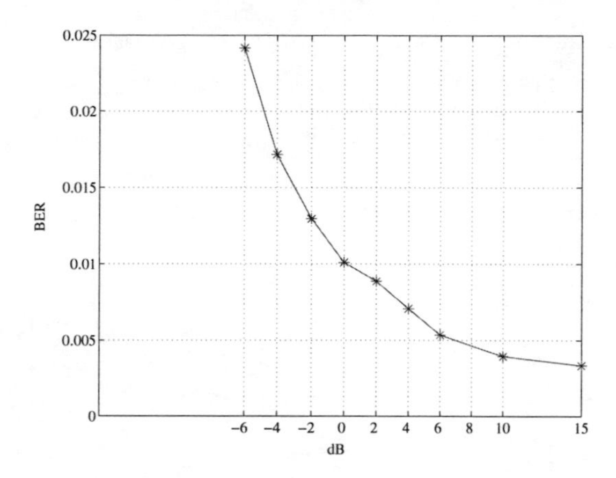

Figure 8.9: Influence of multi-band equalizer on the BER

The intention of this attack is to remove a identification watermark in order to circumvent the copy protection mechanism provided by this marking. To test the robustness 20 different audio files all carrying the same watermark were used and the BER measured.
A second form of this attack is to estimate the original by averaging of the same watermarked audio tracks carrying different watermarks (see equation (8.5) on page 158). This is critical in situations where the watermark carries information of different customers for identification purposes (see section 3.4 on page 46). The 10 different watermarked copies of each original were used in order to test the robustness. The results in table 8.6 verify the robustness against col-

Collusion Attack	*BER*
Approximating the original	0.29
Approximating the watermark	8×10^{-3}

Table 8.6: Collusion attacks and the corresponding BER

lusion attacks trying to approximate the watermark. In contrast the

high BER in the case of the approximation of the original demonstrate the success of this kind of attack.
All the audio files used in the tests so far have been short segments (see table 8.4 on page 177) from longer audio tracks. Having the

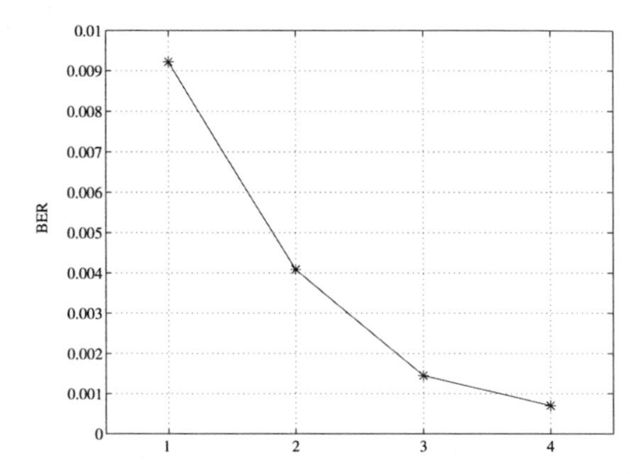

Figure 8.10: Influence of the number of embedded watermarks on the BER

whole watermarked audio file should increase the robustness of the watermark due to the higher redundancy achieved by embedding the watermark a multiple times. A last test was performed showing the dependence of the robustness of the watermark on the length of the segment by evaluating the BER as a function of the number of watermarks embedded (see figure 8.10). The results verify the considerable decrease of the BER by an order of magnitude if the watermark is embedded four times in the audio track.

8.3.4 Robustness against Desynchronization Attacks

Desynchronization attacks like the warping of the time axes presented in section 8.2.3 can be easily implemented even by people with a modest skill in signal processing. To test the robustness against the time-warp attack performed during the SDMI challenge the parameter p determining the period of the warp via

$$T = \frac{2}{p} \quad [sec] \tag{8.33}$$

can be fixed to a value of 2 leading to a period T of 1 second. The maximum number of samples $4d$ delayed is used as attack strength. The results of varying the number of delayed samples are shown in figure 8.11.

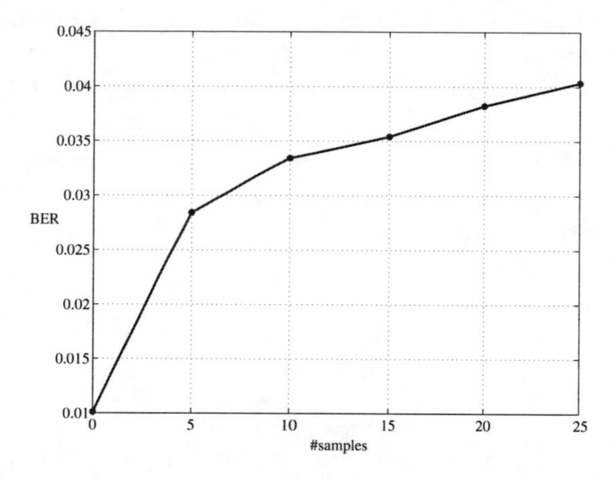

Figure 8.11: Influence of the number of samples on the BER.

Besides the principle vulnerability of the algorithm against time and frequency scaling manipulations, the results verify a moderate robustness against a time warping of the watermarked audio tracks.

8.3.5 Robustness against Embedding Attacks

An obvious embedding attack is to embed the own watermark into the audio track in order to claim ownership if the attacker has access to both the encoder and decoder of the watermarking system. In this context two questions have to addressed:

- Can both watermarks be decoded independently?
- What is the influence of the overmarking on the robustness respectively the probability for false detection on the two watermarks?

Both questions can be answered by measuring the BER of the two watermarks embedded. In this case the second watermark and the associated key are different from the first one, since using the same watermark makes no sense in the context of this attack. Moreover the probability that the attacker generates the same pattern is very low because of the huge number of possible patterns. For $M = 384$ the number of different patterns $N_{\mathbf{pn}} = \binom{2M}{M}$ is approximately $4.86 \cdot 10^{181}$. The test performed shows the BER for the two watermarks with $\mathbf{w_2}$ embedded after $\mathbf{w_1}$ (see table 8.7). Test result showing the BER($\mathbf{w_i}$)

Overmarking	*BER*
Detection of $\mathbf{w_0}$	1.01×10^{-2}
Detection of $\mathbf{w_1}$ out of two	1.13×10^{-2}
Detection of $\mathbf{w_2}$ out of two	1.16×10^{-2}

Table 8.7: Embedding attacks and the corresponding BER

for the two watermarks embedded. The result is averaged over each of the 4000 files for the different watermarks and key combinations. Since the probability of false detection P_F of the different watermarks gives no reliable evidence on the order of embedding the attacker might perform a deadlock attack as presented in section 8.2.4 in order to create a stand-off. A possible countermeasure against this kind of attack is the usage of hash functions calculated from the original audio track as the payload of the watermark. This hash value calculated from the original data set can be used to uniquely identify the order of embedding respectively the copyright owner

of the original audio file. On the other hand using hashes as the watermark require a considerably higher payload size of 160 bit if one uses the Secure Hash Algorithm (SHA) resulting in a longer time segment for embedding one watermark.
In turn the question is if a reduced number of bits from the hash value can be used as a payload size results in a collusion probability equal or less the probability of false detection.
From the properties of the hash function [68] one can easily see that selecting any arbitrary but fixed sequence of bits from the output bits generated by such a function the individual bits selected retain the properties of the entire function, particularly the avalanche effect.
Assuming that m is the number of original audio files, an ideal cryptographic hash function, and a payload size of k bits, the number of compact representations is $N = 2^k$; the ratio of the number of compact representations without collisions to the total number of vectors is $\prod_{i=0}^{m-1}\left(1 - \frac{i}{N}\right)$. Since $1 - x \leq e^{-x}$ (for $x > 0$) one obtains for the probability of no collision

$$\prod_{i=0}^{m-1}\left(1 - \frac{i}{N}\right) \leq \prod_{i=0}^{m-1} e^{-\frac{i}{N}} = e^{-\sum_{i=0}^{m-1}\frac{i}{N}} = e^{-\frac{m(m-1)}{2N}} \tag{8.34}$$

Thus, the probability of a collision after inserting $m = \sqrt{2N}$ keys is $1 - \frac{1}{e}$. For a 32 bit payload this implies a collision once for every 90,000 subjects; for a 64 bit payload one collision in 6 billion subjects can be expected, which may be considered acceptable — hence the reduction from 160 to a 64 bit payload. Obviously, these calculations do not take into account that the payload may not be extracted faithfully; if this is to be avoided, the payload size must be increased further to accommodate error detection or correcting codes, possibly causing audible degradation of the output.
For timestamps which might also be inserted into the payload, the requirements for payload are considerably smaller; one minute resolution can be considered adequate given the presumed purpose of establishing probabilities of culpability; 26 bits are sufficient to encode an entire century at this resolution [22].

8.3.6 Robustness against False Alarms

According to section 8.2.5 producing a false alarm can be accomplished by varying the key K for a fixed object $\mathbf{c}_o$ with the intention to produce a false detection probability P_F above a certain threshold. Besides the attempt to produce a false alarm attack by the variation of the key an attacker can try to take advantage of the knowledge of the published watermarking algorithm. The $2M$ Fourier coefficients used for embedding the different bit pattern are known to the attacker. In turn a first step in the attack would be to construct bit pattern which produce false detection results for each individual bit. What is the probability of producing such kind of a false detection? The probability of false detection for a faked watermark bit can be determined by performing the following steps [7]:

1. Measure the BER as a function of the similarity (B out of M positions match) of generated and embedded pattern.

2. Calculate the probability that $i = B, \ldots, M$ positions match with the embedded bitpattern from:

$$\sum_{i=B}^{M} \frac{\binom{M}{B}\binom{M}{M-B}}{\binom{2M}{M}} \tag{8.35}$$

To measure the detection of a watermark bit with a faked bit pattern, the parameters from the preceding test were used. Afterwards 4 different bit patterns with a certain percentage of similarity with the original bit pattern were constructed and the corresponding detection values for the random variables measured. The result is an average value of the individual z values corresponding to the different bits obtained by evaluating the watermarked audio files.
The results of this test are plotted in figure 8.12 on page 185. To calculate the false alarm probability we use the already high BER for a similarity of 66% with the original watermark. According to equation (8.35), the probability of detecting a watermark which isn't embedded is $\leq 6.29 \times 10^{-13}$. This probability can be further reduced

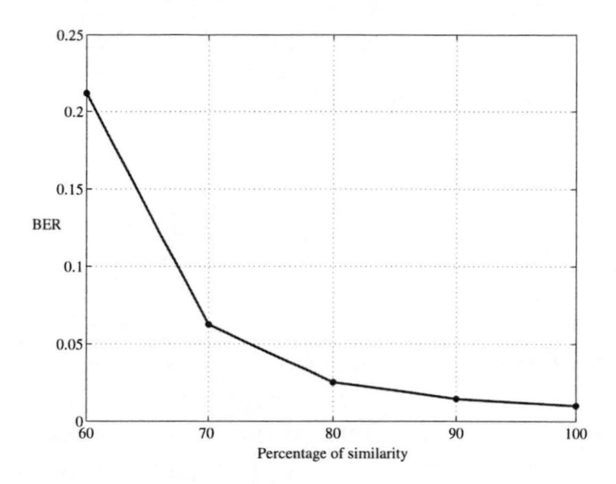

Figure 8.12: Security evaluation about the faked watermarks

by selecting the $2M$ Fourier coefficients in a pseudorandom way by means of the secret key.

8.4 Summary

This chapter investigated the robustness of the watermarks and the security of the underlying watermarking system developed in chapter 6. In a first step an approach for a risk analysis was presented identifying participating classes of actors and their permission to perform watermarking operations grouped into embedding, detecting and removing of the watermarks. The possible attacks for the specific application scenario are then extracted from the so-called attack table presented, which lists the attacks according to the assumptions about the attacker. The attack table includes a categorization of the attacks into three groups with respect to the effect which renders watermarking useless during the detection procedure:

No detection by removing the watermark or misalignment of the watermark and detector.

False detection by embedding of false watermarks.

Unauthorized detection by producing false alarms.

For each class of attack the A©WA algorithm was evaluated. Quantative measures for evaluating the robustness were derived from the experimentally determined bit-error rate (BER). The BER was measured as a function of the attack strength for the removal, desynchronization and embedding attacks. The results show the good robustness of the watermarks embedded via the A©WA algorithm for a variety of different attacks performed. The robustness can further be improved by taking into account detailed knowledge about the intended application and corresponding boundary conditions. This was demonstrated by an approach of circumventing the deadlock attack with the use of hash functions. To investigate the robustness against unauthorized attacks two scenarios were evaluated. In a simulated brute-force attack the secret key used in the detection process was varied in order to produce false alarms. A second kind of attack was performed by taking into account knowledge of the underlying algorithm, leading again to a vanishing probability for falsely detecting the watermark. This verified the robustness of the A©WA algorithm against unauthorized detection attacks.

CHAPTER

NINE

Conclusions

In this thesis, an algorithm for digital watermarking of audio data was designed and implemented, which preserves the perceptual quality of the marked audio data by shaping the embedded signal according to an elaborate psychoacoustic model while at the same time ensuring the robustness of the embedded signal.

The validity of the model as well as the performance of the algorithm developed with regard to the quality of the watermarked signal and the robustness of the embedded signal was shown by means of an exhaustive experimental evaluation.

Moreover, application scenarios for the digital watermarking algorithm were described and subjected to a detailed risk and attack potential analysis. Based on these analytical approaches, countermeasures were identified and described where possible. As a result of this analysis, the suitability of digital audio watermarking mechanisms to individual application scenarios was elaborated in each case.

This chapter provides a reprise and conclusion by presenting an overview of the contributions in section 9.1, the limitations in applying digital watermarking technologies in section 9.2 and possible directions for future research in section 9.3.

9.1 Summary of Contributions

Based on background material on digital watermarking provided in chapter 2 and a description of the psychoacoustic model used in chapter 4, the requirements and hence the design criteria for the watermarking algorithm were elaborated in chapter 6.
Based on these criteria and the psychoacoustic model, the design of an audio watermarking algorithm has been described. This algorithm permits the flexible adjustment of the perceptual quality of the carrier signal as well as the robustness and data rate of the embedded digital watermark signal according to the specific requirements of an application scenario. Moreover, the algorithm described in this dissertation permits the simultaneous embedding of multiple non-interfering watermarks in the same carrier signal.

- A communication-theoretic model was developed which permitted the analysis of both existing watermarking algorithms and the newly developed algorithm. This model allowed the identification of deficiencies in existing approaches and also assisted in identifying possible future research directions (see section 9.3).

- A requirement classification framework was developed which permitted the categorization of multiple, in some cases contradicting, requirements and the derivation of optimal harmonized requirements and parameterizations for specific application scenarios.

- Based on the psychoacoustic model and the requirements classification framework, an algorithm for digital watermarking of audio data has been formulated. For this purpose, a mathematical model for embedding and detection of digital watermarks has been developed, which is independent of the embedding domain and relies on statistical features of the data set.

  It has been shown that the application of this mathematical model to the magnitudes of Fourier coefficients ensures a highly robust digital watermark while at the same time minimizing distortions arising from the use of the original signal's phase.

Furthermore, it has been shown that the robustness metric of the so-called embedding factor developed in this dissertation and the relation to the carrier signal as well as the psychoacoustic features enabled the adjustment between quality and robustness requirements.

- Moreover, it has been shown that the basic algorithm can be generalized to permit the embedding of n bits of information.

  Furthermore, the embedding of several watermarks simultaneously relying only on the orthogonality of the basis functions of the embedding domain used has been demonstrated.

- The approach of integrating a psychoacoustic model based on the calculation of the masking threshold has been validated; the results obtained in this dissertation also extend to psychoacoustic models of higher complexity.

- Methods for evaluating the quality of watermarked audio signals based on subjective listening tests and objective measurements have been described and their applicability has been classified according to the quality constraints imposed by the specific application scenario.

  Based on the evaluation protocol thus developed, both subjective and objective quality evaluations have been conducted and the results summarized in this dissertation.

- A detailed risk analysis based on the application scenario with a description of attacks and possible countermeasures has been described.

9.2 Practical Limits of Digital Watermarking

As discussed in chapter 3, the usage of digital watermarking technologies particularly for copyright and copy protection applications is limited in its efficacy in part by technical considerations. However, particularly for copyright and copy protection applications, legal constraints exist in most jurisdictions, which place further limits on the overall efficacy in this application scenario.

In the case of copyright protection applications, digital watermarks are used to identify the audio material if it is located but cannot assist in identifying the source of this material unless the source's identity is also embedded in a digital watermark. Such additional markings, however, are problematic in that they must either be generated centrally and transmitted to each individual customer, resulting in a significant cost and time overhead, or the markings must be embedded at customer site. Since the equipment for marking (and presumably playback) are under the control of the customer, the possibility of tampering as detailed in [10] cannot be dismissed. As a result, the evidentiary value of such identification markings may be called into question.

Another possible approach would be the use of automatic search engines or search agents trying to locate the origin of unauthorized resold or redistributed material. However, performing the search by central analysis would be difficult in terms of bandwidth and processing power and the use of search agents would assume that end users grants access for executing code on their system, which appears unlikely in the event that such a user indeed harbors unauthorized copies of protected material.

Furthermore, the legal protection afforded against invasions of privacy in civil matters without probable cause appear to limit the potential feasibility of this approach — in addition to considerations regarding the cost of litigation.

Moreover, proper identification of the source (or an individual, respectively) is even more complicated if file sharing services are used for data exchange since these provide a certain level of anonymization. In this case actions may not be clearly associated with an individual even under the standards of civil litigation relying only on the preponderance of evidence.

This situation is different for content providers selling creations and therefore have to expose and clearly identify themselves to the public as well as signal their intent to conduct such sales.
However, litigation may also be difficult even in this case since the content provider may reside in a non-cooperative jurisdiction and hence be shielded from claims brought against him.
In addition, although copy protection has frequently been identified as one of the most desirable deployment scenarios for the music industry (where complete detectors or even embedders and possibly key material are present in the hands of adversaries), this particular application scenario also implies a vulnerability to classes of attacks such as oracle attacks.
These kind of attacks, described in chapter 8, can be classified as protocol-specific or usage-specific and hence must be dealt with by way of careful design of protection mechanisms and application scenarios.

9.3 Future Research

Besides the constraints particular to copyright and copy protection applications imposed by legal considerations as well as cost (i.e. balancing the cost of embedding effective tamper-resistant marking and content protection mechanisms against the potential losses incurred without protective measures), for digital watermarks to be useful in content protection application scenarios, research is still required particularly in the area of increasing robustness to the various types of attacks described in chapter 8.
There exists the need for additional research in digital watermarking particularly with regard to improved robustness against the various types of desynchronization attacks. Such robustness can presumably be achieved if more information (i.e. signal-specific feature-level

semantics) can be used for registration of signals that have been subjected to manipulation or outright deliberate desynchronization attacks.
Similar improvements can also be conjectured for signal processing attacks if higher level semantics are considered based on the significant improvements in robustness achieved in the past with the introduction of advanced perceptual modeling (see section 6.5 on page 105) into digital watermarking.
Moreover, for some application scenarios discussed here such as monitoring applications, any such improved signal registration and detection algorithm must be limited in its computational complexity so that the detection of a watermark (and hence of a license or other copyright violation) can be accomplished in real time and with commensurate expenses for equipment in the case of audio data. Obviously, this constraint of computational complexity is less severe in the case of offline detection and verification.
As it is the case for any content protection mechanism the usage and integration of digital watermarking technologies — and indeed multiple redundant or integrated mechanisms are highly desirable — must be balanced carefully between multiple conflicting requirements such as the initial and per-unit cost of mechanism, the extent to which customers may be inconvenienced and thus kept from purchasing either the product or subsequent offerings, and the actual protection.
However, the protection of intellectual property in an increasingly digital domain with abundant computational capabilities and bandwidth capacity will remain a multidimensional challenge with technical, economical, and legal aspects demanding equal careful attention.

GLOSSARY

Bark A non-SI unit of measurement named after the physicist Barkhausen modeling the representation of sound in the human auditory system more closely than SI units. The Bark scale assumes that one unit corresponds to a constant length (1.3 mm) along the basilar membrane. The Bark frequency scale ranges from 1 to 24 consisting of center frequencies and band edges to be interpreted as samplings of a continuous variation in the frequency response of the ear to a sinusoid or narrowband noise signal, corresponding to the first 24 critical bands of hearing and range up to 15.5 kHz, implying that the highest sampling rate for which the Bark scale is defined up to the Nyquist limit is 31 kHz.

Cepstrum Given a sequence $x(n)$ with a z-Transform of $X(z)$, the complex cepstrum is defined as the inverse z-Transform of the natural logarithm of the z-Transform of $x(n)$ or $c_x(n) = Z^{-1}\{\ln Z\{x(n)\}\} = Z^{-1}\{C_x(z)\}$. If the complex cepstrum exists, $C_x(w) = \ln X(w) = \sum_{n-\infty}^{\infty} c_x(n)e^{-j\omega n}$ converges at the unit circle and $c_x(n)$ is obtained from the IFT of $\ln X(w)$. Expressing $X(w)$ in terms of magnitude and phase, the complex cepstrum can also be separated into magnitude and phase. In audio processing, only the real component $c_x(n) = \frac{1}{2\pi}\int_{-\pi}^{\pi} \ln|X(w)|e^{jn\omega}d\omega$ is used.

Cryptography From the Greek κρυπτός (hidden, secret) and γράφειν (to write), literally "secret writing".

EBU European Broadcasting Union.

Multisession CD A multisession CD is a recordable CD format (like a CD-R) that allows the recording of a compact disk to be conducted in more than one recording session. If there is free space left on the CD after the first session, additional data can be written to it at a later date. Each session has its own lead in, program area, and lead out.

Pseudo Noise A digital signal with noise-like properties.

Secure Digital Music Initiative The Secure Digital Music Initiative (SDMI) consists of a consortium of music companies trying to standardize technologies for the protection of digital music with regard to playing, storing and distributing.

Steganography From the Greek στεγᾰ́νός (closely covered, sheathed) and γράφειν (to write), literally "hidden writing".

LIST OF ACRONYMS

AAC .... Advanced Audio Coding.

ACR........... Absolute Category Rating.

A©WA .... Audio ©opyright protection by WAtermarking.

ASCAP........ The American Society of Composers, Artists and Publishers.

BPM .... Blind Pattern Matching.

CD .... Compact Disc.

CQS........... Continuous Quality Scale.

DVD CCA..... DVD Copy Control Association.

EBU .... European Broadcasting Union.

FFT .... Fast Fourier Transformation.

FIR............ Finite Impulse Response.

FTP .... File Transfer Protocol.

IFPI .... International Federation of the Phonographic Industry.

IM .... Instant Messaging.

IPR .... Intellectual Property Right.

IRC .... Internet Relay Chat.

ITU-R .... Radiocommunication Sector of the ITU.

LSB .... Least-Significant-Bit.

MCLT .... Modulated Complex Lapped Transform.

MPEG .... Motion Picture Expert Group.

MUSHRA ..... Multi Stimulus with Hidden Reference Anchors.

ODG .... Objective Difference Grade.

PAM .... Psychoacoustic model.

PCM .... Pulse Code Modulation.

PEAQ .... Perceived Audio Quality.

PN .... Pseudo Noise.

PRNG .... Pseudo Random Noise Generator.

PSC .... Power Spectrum Condition.

RMS .... Root Mean Square.

SACD .... Super Audio CD.

SDG .... Subjective Difference Grade.

SDMI .... Secure Digital Music Initiative.

SHA .... Secure Hash Algorithm.

SMR .... Signal-to-Mask-Ratio.

SVCD .... Super Video CD.

WMS .... Watermarking Minimum Segment.

Bibliography

[1] Adrian Perrig and Andrew Willmott. Digital Image Watermarking in the "Real World". Tech. rep., Carnegie Mellon University Computer Science Department, Pittsburgh, PA, USA, Jan. 1998.

[2] Arnold, M. Audio Watermarking: Features, Applications and Algorithms. In *Proceedings of the IEEE International Conference on Multimedia and Expo (ICME 2000)* (New York, NY, USA, July 2000), IEEE Press, pp. 1013–1016.

[3] Arnold, M. Subjective and Objective Quality Evaluation of Watermarked Audio Tracks. In *Proceedings of Second International Conference on WEB Delivering of Music (WEDELMUSIC 2002)* (Darmstadt, HE, Germany, Dec. 2002), C. Busch, M. Arnold, P. Nesi, and M. Schmucker, Eds., vol. 4675, IEEE Computer Society Press, pp. 161–167.

[4] Arnold, M. Attacks on Digital Audio Watermarks and Countermeasures. In *Proceedings of Third International Conference on WEB Delivering of Music (WEDELMUSIC 2003)* (Leeds, LS2 9JT, United Kingdom, Sept. 2003), K. Ng, C. Busch, and P. Nesi, Eds., IEEE Computer Society Press, pp. 55–62.

[5] Arnold, M., and Huang, Z. Blind Detection of Multiple Audio Watermarks. In *Proceedings of First International Conference on WEB Delivering of Music (Wedelmusic 2001), Florence, Italy* (Florence, Italy, Nov. 2001), P. Nesi, Pierfrancesco, and C. Busch, Eds., IEEE Computer Society, pp. 4–11.

[6] Arnold, M., and Huang, Z. Fast Audio Watermarking Concepts and Realizations. In *Proceedings of Electronic Imaging 2004, Security and Watermarking of Multimedia Contents VI* (San Jose, CA, USA, Jan. 2004), P. W. Wong and E. J. Delp, Eds., SPIE, p. To appear.

[7] Arnold, M., and Kanka, S. MP3 Robust Audio Watermarking. In *DFG V III D II Watermarking Workshop* (Erlangen, Germany, Oct. 1999).

[8] Arnold, M., and Lobisch, O. Real-time Concepts for Block-based Watermarking Schemes. In *Proceedings of Second International Conference on WEB Delivering of Music (WEDELMUSIC 2002)* (Darmstadt, HE, Germany, Dec. 2002), C. Busch, M. Arnold, P. Nesi, and M. Schmucker, Eds., vol. 4675, IEEE Computer Society Press, pp. 156–160.

[9] Arnold, M., and Schilz, K. Quality Evaluation of Watermarked Audio Tracks. In *Proceedings of Electronic Imaging 2002, Security and Watermarking of Multimedia Contents IV* (San Jose, CA, USA, Jan. 2002), P. W. Wong and E. J. Delp, Eds., vol. 4675, SPIE, pp. 91–101.

[10] Arnold, M., Schmucker, M., and Wolthusen, S. *Techniques and Applications of Digital Watermarking and Content Protection.* Artech House, Boston, MA, USA, 2003.

[11] Barni, M., Bartolini, F., Cappellini, V., and Piva, A. A DCT-Domain System for Robust Image Watermarking. *Signal Processing 66*, 3 (May 1998), 357–372.

[12] Bassia, P., and Pitas, I. Robust Audio Watermarking in the Time Domain. In *Signal Processing IX, Theories and Applications: Proceedings of EUSIPCO-98, Ninth European Signal Processing Conference* (Sept. 1998), S. T. et al., Ed., Typorama Editions, pp. 25–28.

[13] Bauer, F. L. *Kryptologie: Methoden und Maximen*, 2nd ed. Springer-Verlag, Heidelberg, Germany, 1991.

[14] Bender, W., Gruhl, D., Morimoto, N., and Lu, A. Techniques for Data Hiding. *IBM Systems Journal 35*, 3 & 4 (1996), 313–336.

[15] Biddle, P., England, P., Peinado, M., and Willman, B. The Darknet and The Future of Content Distribution. In *2002 ACM Workshop on Digital Rights Management* (Washington DC, USA, Jan. 2003), N. of editor, Ed., vol. 000 of *Lecture Notes in Computer Science*, Springer-Verlag, pp. –.

[16] Boeuf, J., and Stern, J. P. An Analysis of One of the SDMI Candidates. In *Information Hiding: 4th International Workshop* (Pittsburgh, PA, USA, 2001), I. S. Moskowitz, Ed., vol. 2137 of *Lecture Notes in Computer Science*, Springer-Verlag, pp. 395–409.

[17] Boneh, D., and Shaw, J. Collusion-Secure Fingerprinting for Digital Data. In *Proceedings of Advances in Cryptology, CRYPTO '95* (Santa Barbara, CA, USA, Aug. 1995), D. Coppersmith, Ed., vol. 963 of *Lecture Notes in Computer Science*, Springer-Verlag, pp. 452–465.

[18] Boney, L., Tewfik, A. H., and Hamdy, K. N. Digital Watermarks for Audio Signals. In *IEEE International Conference on Multimedia Computing and Systems* (Hiroshima, Japan, June 1996), IEEE, Ed., IEEE Press, pp. 473–480.

[19] Bosi, M., Brandenburg, K., Quackenbush, S., Fielder, L., Akagiri, K., Fuchs, H., Dietz, M., Herre, J., Davidson, G., and Oikawa, Y. ISO/IEC MPEG-2 Advanced Audio Coding. *Journal of the Audio Engineering Society 45*, 10 (Oct. 1997), 789–814.

[20] Brandenburg, K., and Sporer, T. Evaluation of quality using perceptual criteria. In *Proceedings of the 11th International AES Conference on Audio Test and Measurement* (Portland, OR, USA, May 1992), pp. 169–179.

[21] Busch, C., Funk, W., and Wolthusen, S. Digital Watermarking: From Concepts to Real-Time Video Applications. *IEEE Computer Graphics and Applications 19*, 1 (Jan. 1999), 25–35.

[22] Busch, C., and Wolthusen, S. Tracing Data Diffusion in Industrial Research with Robust Watermarking. In *Proceedings*

of the 2001 Fourth Workshop on Multimedia Signal Processing (MMSP'01) (Cannes, France, Oct. 2001), J.-L. Dugelay and K. Rose, Eds., IEEE Press, pp. 207–212.

[23] CHENG, S., YU, H., AND XIONG, Z. Enhanced Spread Spectrum Watermarking of MPEG-2 AAC Audio. In *IEEE International Conference on Acoustics, Speech, and Signal Processing (ICASSP)* (May 2002), vol. 4, IEEE Press, pp. 3728–3731.

[24] COSTA, M. Writing on Dirty Paper. *IEEE Transactions on Information Theory 29*, 3 (May 1983), 439–441.

[25] COX, I. J., KILIAN, J., LEIGHTON, T., AND SHAMOON, T. Secure Spread Spectrum Watermarking for Multimedia. Tech. Rep. 95-10, NEC Research Institute, 1995.

[26] COX, I. J., AND LINNARTZ, J.-P. M. G. Some General Methods for Tampering with Watermarks. *IEEE Journal on Selected Areas in Communications 16*, 4 (May 1998), 587–593.

[27] COX, I. J., MILLER, M. L., AND BLOOM, J. A. *Digital Watermarking*. The Morgan Kaufmann Series in Multimedia Information and Systems. Morgan Kaufmann Publishers, San Francisco, CA, USA, 2002.

[28] CRAVER, S., MEMON, N., YEO, B.-L., AND YEUNG, M. Can Invisible Watermarks Resolve Rightful Ownerships? Tech. Rep. 20509, IBM Research Divison, Yorktown Heights, NJ, USA, July 1996.

[29] CRAVER, S., YEO, B.-L., AND YEUNG, M. Technical Trials and Legal Tribulations. *Communications of the Association for Computing Machinery 41*, 7 (July 1998), 45– 54.

[30] CRAVER, S. A., WU, M., LIU, B., STUBBLEFIELD, A., SWARTZLANDER, B., WALLACH, D. S., DEAN, D., AND FELTEN, E. W. Reading Between the Lines: Lessons from the SDMI Challenge. In *Proceedings of the 10th USENIX Security Symposium* (Washington D.C., USA, Aug. 2001).

[31] de Vigenere, B. *Traicté des chiffres, et secretes manieres d'escrire.* Paris, France, 1586.

[32] Dittmann, J. *Sicherheit in Medienströmen: Digitale Wasserzeichen.* PhD thesis, TU Darmstadt, Darmstadt, Germany, 1999.

[33] Doyle, A. C. The Adventure of the Dancing Men. *Strand Magazine 26*, 156 (Dec. 1903).

[34] DVD Copy Control Association. Request for Expressions of Interest. Tech. rep., DVD Copy Control Association, Apr. 2001.

[35] EBU Project Group B/AIM. EBU Report on the Subjective Listening Tests of Some Commercial Internet Audio Codecs. Tech. rep., European Broadcasting Union (EBU), June 2000.

[36] Gardner, M. *Codes, Ciphers and Secret Writing.* Dover Publications, Mineola, NY USA, 1972.

[37] Gruhl, D., Lu, A., and Bender, W. Echo Hiding. In *Information Hiding: First International Workshop* (Cambridge, UK, May 1996), R. J. Anderson, Ed., vol. 1174 of *Lecture Notes in Computer Science*, Springer-Verlag, pp. 295–315.

[38] Haitsma, J., van der Veen, M., Kalker, T., and Bruekers, F. Audio Watermarking for Monitoring and Copy Protection. In *Proceedings of the ACM Multimedia 2000 Workshop* (Los Angeles, CA, USA, Nov. 2000), ACM Press, pp. 119–122.

[39] Hartung, F., and Girod, B. Digital Watermarking of Uncompressed and Compressed Video. *Signal Processing 66*, 3 (May 1998), 283–301.

[40] Hartung, F., Su, J. K., and Girod, B. Spread Spectrum Watermarking: Malicious Attacks and Counterattacks. In *International Conference on Security and Watermarking of Multimedia Contents* (San Jose, CA, USA, Jan. 1999), Ping Wah Wong and Edward J. Delp, Ed., SPIE, pp. 147–158.

[41] Hayhurst, J. D. The Pigeon Post into Paris 1870–1871. Privately published by the author; Dewey: 383.144 H331p at University of Texas at Austin Library, 1970.

[42] Holt, L., Maufe, B. G., and Wiener, A. Encoded Marking of a Recording Signal. U.K. Patent GB 2196167., Mar. 1988. Granted January 1991, lapsed May 1995.

[43] International Federation of the Phonographic Industry. Request for Proposals – Embedded Signalling Systems. Tech. rep., International Federation of the Phonographic Industry, London, UK, 1997.

[44] International Federation of the Phonographic Industry. IFPI Music Piracy Report, June 2001.

[45] International Telecommunication Union. *Subjective Assessment of Sound Quality*, 1990.

[46] International Telecommunication Union. *Methods for Subjective Assessement of Small Impairments in Audio Systems including Multichannel Sound Systems*, 1997.

[47] International Telecommunication Union. *A Method for Subjective Listening Tests for Intermediate Audio Quality – Contribution from the EBU to ITU Working Party 10-11Q*, 1998.

[48] International Telecommunication Union. *Method for Objective Measurements of Perceived Audio Quality (PEAQ)*, 1998.

[49] ISO/IEC Joint Technical Committee 1 Subcommittee 29 Working Group 11. Information technology - Coding of moving pictures and associated audio for digital storage media at up to about 1.5Mbit/s Part 3: Audio. ISO/IEC 11172-3, 1993.

[50] ISO/IEC Joint Technical Committee 1 Subcommittee 29 Working Group 11: Coding of Moving Pictures and Audio. *MPEG-2 Advanced Audio Coding, AAC*, 1997.

[51] Johnson, N., and Jajodia, S. Steganography: Seeing the Unseen. *IEEE Computer 31*, 2 (Feb. 1998), 26–34.

[52] Kahn, D. *The Codebreakers: The Comprehensive History of Secret Communication from Ancient Times to the Internet*, 2nd ed. Scribner, New York, NY, USA, 1996.

[53] Kalker, T., Linnartz, J.-P. M. G., and van Dijk, M. Watermark Estimation through Detector Analysis. In *Proceedings of the International Conference on Image Processing* (Oct. 1998), IEEE Press, pp. 425–429.

[54] Katzenbeisser, S., and Petitcolas, F. A. P., Eds. *Information Hiding: Techniques for Steganography and Digital Watermarking*. Artech House, Boston, MA, USA, 2000.

[55] Kerkhoffs, A. La Cryptographie Militaire. *Journal des Sciences Militaires 9th series* (Jan./Feb. 1883), 5–38,161–191.

[56] Kirovski, D., and Malvar, H. Robust Covert Communication over a Public Audio Channel Using Spread Spectrum. In *Information Hiding: 4th International Workshop* (Portland, OR, USA, Apr. 2001), I. S. Moskowitz, Ed., vol. 2137 of *Lecture Notes in Computer Science*, Springer-Verlag, pp. 354–368.

[57] Kirovski, D., and Malvar, H. Robust Spread-Spectrum Audio Watermarking. In *IEEE International Conference on Acoustics, Speech and Signal Processing (ICASSP)* (Salt Lake City, UT, USA, May 2001), IEEE Press, pp. 1345–1348.

[58] Ko, B.-S., Nishimura, R., and Suzuki, Y. Time-Spread Echo Method for Digital Audio Watermarking using PN Sequences. In *International Conference on Acoustics, Speech and Signal Processing (ICASSP)* (Orlando, FL, USA, May 2002), IEEE Press, pp. 2001–2004.

[59] Koch, E., and Zhao, J. Towards Robust and Hidden Image Copyright Labeling. In *Proceedings of 1995 IEEE Workshop on*

Nonlinear Signal and Image Processing (June 1995), I. Pitas, Ed., IEEE Press, pp. 452–455.

[60] KUO, S.-S., JOHNSTON, J., TURIN, W., AND S.R., Q. Covert Audio Watermarking using Perceptually Tuned Signal Independent Multi]band Phase Modulation. In *IEEE International Conference on Acoustics, Speech, and Signal Processing (ICASSP)* (May 2002), vol. 2, IEEE Press, pp. 1753–1756.

[61] KUTTER, M., VOLOSHYNOVSKIY, S., AND HERRIGEL, A. The Watermark Copy Attack. In *Proceedings of Electronic Imaging 2000, Security and Watermarking of Multimedia Contents II* (Jan. 2000), P. W. Wong and E. J. Delp, Eds., SPIE, pp. 371–381.

[62] LACY, J., QUACKENBUSH, S. R., REIBMAN, A. R., AND SNYDER, J. H. Intellectual Property Protection Systems and Digital Watermarking. *Optics Express 3*, 12 (Dec. 1998), 478–484.

[63] LANGELAAR, G. C., LAGENDIJK, R. L., AND BIEMOND, J. Removing Spatial Spread Spectrum Watermarks by Non-linear filtering. In *Ninth European Signal Processing Conference* (Island of Rhodos, Greece, Sept. 1998), pp. 2281–2284.

[64] LEHN, J., AND WEGMANN, H. *Einführung in die Statistik*. B.G. Teubner, Stuttgart, Deutschland, 1985.

[65] LEVY, S. The Big Secret. *Newsweek* (July 2002).

[66] LIN, S., AND COSTELLO, JR., D. J., Eds. *Error Control Coding: Fundamentals and Applications*. Prentice-Hall Series in Computer Applications in Electrical Engineering. Prentice Hall, Englewood Cliffs, NJ, USA, 1983.

[67] LINNARTZ, J.-P. M. G., AND VAN DIJK, M. Analysis of the Sensitivity Attack against Electronic Watermarks in Images. In *Information Hiding: Second International Workshop* (Portland, OR, USA, 1998), D. Aucsmith, Ed., vol. 1525 of *Lecture Notes in Computer Science*, Springer-Verlag, pp. 258–272.

[68] LUBY, M. *Pseudorandomness and Cryptographic Applications*. Princeton Computer Science Notes. Princeton University Press, Princeton, NJ, USA, 1996.

[69] MANNOS, J. L., AND SAKRISON, D. J. The Effects of a Visual Criterion on the Encoding of Images. *IEEE Transactions on Information Theory IT-20*, 4 (July 1974), 525–536.

[70] NAHRSTEDT, K., AND QIAO, L. Non-Invertible Watermarking Methods for MPEG Video and Audio. In *Multimedia and Security Workshop at ACM Multimedia 98* (Bristol, UK, Sept. 1998), J. Dittmann, P. Wohlmacher, P. Horster, and R. Steinmetz, Eds., GMD – Forschungszentrum Informationstechnik GmbH, pp. 93–98.

[71] NEUBAUER, C. Digitale Wasserzeichen für unkomprimierte und komprimierte Audiodaten. In *Sicherheit in Netzen und Medienströmen* (Heidelberg, Germany, Sept. 2000), R. S. Markus Schumacher, Ed., Springer-Verlag, pp. 149–158.

[72] NEUBAUER, C., HERRE, J., AND BRANDENBURG, K. Continuous Steganographic Data Transmission Using Uncompressed Audio. In *Information Hiding: Second International Workshop* (Portland, OR, USA, 1998), D. Aucsmith, Ed., vol. 1525 of *Lecture Notes in Computer Science*, Springer-Verlag, pp. 208–217.

[73] NOLL, P. MPEG Digital Audio Coding. *IEEE Signal Processing Magazine 9* (Sept. 1997), 59–81.

[74] NORMAN, B. *Secret Warfare*, 2nd ed. Acropolis Books, Washington D.C., USA, 1996.

[75] OH, H., SEOK, J., HONG, J., AND YOUN, D. New Echo Embedding Technique for Robust and Imperceptible Audio Watermarking. In *International Conference on Acoustics, Speech and Signal Processing (ICASSP)* (Orlando, FL, USA, 2001), IEEE Press, pp. 1341–1344.

[76] Pan, D. A Tutorial on MPEG/Audio Compression. *IEEE Multimedia 2*, 2 (1995), 60–74.

[77] Perrig, A. *A Copyright Protection Environment for Digital Images*. PhD thesis, École Polytechnique Fédérale de Lausanne, Lausanne, Switzerland, 1997.

[78] Petitcolas, F. A. P., Anderson, R. J., and Kuhn, M. G. Attacks on Copyright Marking Systems. In *Information Hiding: Second International Workshop* (Portland, OR, USA, Apr. 1998), D. Aucsmith, Ed., vol. 1525 of *Lecture Notes in Computer Science*, Springer-Verlag, pp. 218–238.

[79] Petitcolas, F. A. P., and Kirovski, D. The Blind Pattern Matching Attack on Watermark Systems. In *Proceedings 2002 IEEE International Conference on Acoustics, Speech, and Signal Processing (ICASSP)* (Orlando, FL, USA, May 2002), IEEE Press, pp. 3740–3743.

[80] Piron, L., Arnold, M., Kutter, M., Funk, W., Boucqueau, J.-M., and Craven, F. OCTALIS Benchmarking: Comparison of Four Watermarking Techniques. In *Proceedings of Electronic Imaging '99, Security and Watermarking of Multimedia Contents* (San Jose, CA, USA, Jan. 1999), P. W. Wong and E. J. Delp, Eds., SPIE, pp. 240–250.

[81] Podilchuk, C. I., and Zeng, W. Image-Adaptive Watermarking Using Visual Models. *IEEE Journal on Selected Areas in Communications 16*, 4 (May 1998), 525–539.

[82] Polmar, N., and Allen, T. B. *The Encyclopaedia of Espionage*, 2nd ed. Random House, New York, NY, USA, 1998.

[83] Precoda, K., and Meng, T. Listener Differences in Audio Compression Evaluations. *Audio Engineering Society 45*, 9 (Sept. 1997), 708–715.

[84] Press, W. H., Teukolsky, S. A., Vetterling, W. T., and Flannery, B. P. *Numerical Recipes in C: The Art of Scientific Computing (2nd ed.).* Cambridge University Press, Cambridge, UK, 1992.

[85] Proakis, J. G., and Manolakis, D. M. *Digital Signal Processing: Principles, Algorithms and Applications*, 2nd ed. Macmillan Publishing Company, Basingstoke, UK, 1992.

[86] Roth, V., and Arnold, M. Improved Key Management for Digital Watermark Monitoring. In *Proceedings of Electronic Imaging 2002, Security and Watermarking of Multimedia Contents IV* (San Jose, CA, USA, Jan. 2002), P. W. Wong and E. J. Delp, Eds., vol. 4675, SPIE, pp. 652–658.

[87] Safford, L. F. Statement Regarding Winds Message by Captain L.F. Safford, U.S. Navy, Before the Joint Committee on the Investigation of the Pearl Harbor Attack. Collection of Papers Related to the "Winds Execute" Message, U.S. Navy, 1945; document SRH-210., Jan. 1946. Located in Record Group 457 at the National Archives and Records Administration, College Park, MD, USA.

[88] Schott, C. *Schola steganographia.* Jobus Hertz, 1665.

[89] Secure Digital Music Initiative. SDMI Portable Device Specification. Tech. rep., Secure Digital Music Initiative, July 1999. Document number PDWG99070802, Part 1, Version 1.0 .

[90] Shannon, C. E. A Mathematical Theory of Communication. *Bell System Technical Journal* 27, 3 (July/Oct. 1948), 379–423,623–656. Reprinted as [91].

[91] Shannon, C. E. A Mathematical Theory of Communication. In *Claude Elwood Shannon: Collected Papers*, N. J. A. Shloane and A. D. Wyner, Eds. IEEE Press, Piscataway, NJ, USA, 1993, pp. 5–83.

[92] Shlien, S. Guide to MPEG-1 Audio Standard. In *IEEE Transactions on Broadcasting* (Dec. 1994), vol. 40, pp. 206–218.

[93] Shlien, S., and Soulodre, G. Measuring the Characteristics of "expert" Listeners. In *Proceedings 101st Convention Audio Engineering Society* (Nov. 1996), Audio Engineering Society.

[94] Smend, F. *Johann Sebastian Bach bei seinem Namen gerufen: Eine Noteninschrift und ihre Deutung*. Bärenreiter, Basel, Switzerland, 1950.

[95] Sporer, T. Evaluating Small Impairments with the Mean Opinion Scale – Reliable or just a Guess? In *Proceedings 101st Convention Audio Engineering Society* (Nov. 1996).

[96] Su, J. K., and Girod, B. Power-spectrum Condition for Energy-efficient Watermarking. In *International Conference on Image Processing (ICIP-99)* (Kobe, Japan, Oct. 1999), IEEE Press, pp. 301–305.

[97] Swanson, M. D., Zhu, B., Tewfik, A. H., and Boney, L. Robust Audio Watermarking using Perceptual Masking. *Signal Processing 66*, 3 (May 1998), 337–355.

[98] Tatlow, R. *Bach and the Riddle of the Number Alphabet.* Cambridge University Press, Cambridge, UK, 1991.

[99] Technical Centre of the European Broadcasting Union. *Sound Quality Assessment Material Recordings for Subjective Tests.* Avenue Albert Lancaster, 32, B-1180 Bruxelles (Belgium), Apr. 1988.

[100] Terhardt, E. Calculating Virtual Pitch. *Hearing Research 1* (1979), 155–182.

[101] Tewfik, A. H., and Mansour, M. F. Secure Watermark Detection with Non-Parametric Decision Boundaries. In *Proceedings of the 2002 IEEE International Conference on Acoustics, Speech, and Signal Processing (ICASSP)* (Orlando, FL, USA, May 2002), IEEE Press, pp. 2089–2092.

[102] TRITHEMIUS, J. *Polygraphiae libri sex... Accessit clavis polygraphiae liber unus, eodem authore.* Ioannes Birckmannus & Theodorus Baumius, 1586. Manuscript was finished in 1508; the work was not published until 1571. Copies are located at the National Cryptologic Museum, Ft. George G. Meade, MD, USA and the Hill Monastic Manuscript Library, Bush Center, Saint John's University, Collegeville, MN, USA.

[103] TRITHEMIUS, J. *Steganographia.* Christophorus Küchlerus, 1676. Unfinished manuscript, ended in 1500 but was not published until 1606. A copy is located at the Hill Monastic Manuscript Library, Bush Center, Saint John's University, Collegeville, MN, USA.

[104] VOLOSHYNOVSKIY, S., PEREIRA, S., HERRIGEL, A., BAUMGÄRTNER, N., AND PUN, T. Generalized Watermarking Attack Based on Watermark Estimation and Perceptual Remodulation. In *Proceedings of Electronic Imaging 2000, Security and Watermarking of Multimedia Contents II* (San Jose, CA, USA, Jan. 2000), P. W. Wong and E. J. Delp, Eds., SPIE.

[105] WATSON, A. B. DCT Quantization Matrices Visually Optimized for Individual Images. In *Proceedings of Human Vision, Visual Processing, and Digital Display IV* (San Jose, CA, USA, Feb. 1993), J. Allebach and B. Rogowitz, Eds., SPIE, pp. 202–216.

[106] WINOGRAD, R. P. J., JEMILI, K., AND METOIS, E. Data Hiding within Audio Signals. In *4th International Conference on Telecommunications in Modern Satellite, Cable and Broadcasting Service* (Nis, Yugoslavia, Oct. 1999), pp. 88–95.

[107] ZHAO, J., AND LUO, C. Digital Watermark Mobile Agents. In *Proceedings of NISSC'99* (Arlington, VA, USA, Oct. 1999), pp. 138–146.

[108] ZÖLZER, U. *Digitale Audiosignalverarbeitung*, 2nd ed. B. G. Teubner, Wiesbaden, Germany, 1997.

[109] Zwicker, E., and Fastl, H. *Psychoacoustics: Facts and Models*, 2nd ed. Springer-Verlag, Heidelberg, Germany, 1999.

INDEX